UDC 004+007](477)(091)"1950/..."=111
H40

H40 Volodymyr Nevzorov, Victoria Ugryumova. Innovation in Isolation. The Story of Ukrainian IT from the 1940s to the Present. — Kyiv: IST Publishing 2024. — 224 p.
ISBN 978-617-7948-30-7

The complete history of IT in Ukraine in one book, from the first computer to the present-day companies you know and love. Here you'll find dozens of rare photos from the early days of computing, along with exclusive interviews with Ukrainian tech pioneers past and present.
A compelling and detailed tour of Europe's fastest-growing tech hub, this book is required reading for anyone interested in modern Ukraine.

Photos

GS Pschenichny Central State Film and Photo Archive of Ukraine: 006-010, 021, 027, 029, 032-038, 043, 046, 049 (both), 051, 054, 056, 057-058, 060, 061 (top), 062-064, 067-071, 072 (lower), 073 (upper), 074-077, 080-084, 088-089, 091-095, 105, 124, 126, 132 (right), 134, 136, 138 (upper)

Ihor Borysov: 028, 052-054, 096, 099, 103 (both), 104, 109 (both), 110, 118, 121 (lower), 122, 123, 125-126, 129, 130, 132, 135, 145, 146 (both)

Volodymyr Falin: 015, 085-086, 137 (lower right), 138 (upper), 140-141, 142 (both), 143-144, 145 (left)

Courtesy of Vira Glushkova: 019, 022, 025, 026, 042, 044-045

Semen Grozyan (photographed at Igor Sikorsky Kyiv State Polytechnic Museum): 039, 072 (upper), 073 (lower), 078-079, 112

Mariupol 8-bit Museum: 100 (upper), 120 (both), 121

Courtesy of Volodymyr Puida: 113, 114, 116, 117

Volodymyr Nevzorov: 023, 040 (top), 055

Oleksandr Strelnikov: 110 (upper), 117

Yakov Khalip: 138 (lower left)

Museum of Development of Computer Science and Technologies in Ukraine: 018, 040, 061 (both lower)

———ist publishing

ISBN 978-617-7948-30-7

Authors:
Volodymyr Nevzorov
Victoria Ugryumova
Commissioned by:
MacPaw Inc.
Translation to English by:
Hanna Leliv
Reilly Costigan-Humes
Editor of the English text:
Rachel McVey
Idea by:
Yuliia Petryk
Project coordinators:
Alyona Gorbatko
Ievgenii Kalnyk
Nina Bohush
Oleksandra Lytvynenko
Art director:
Aliona Solomadina
Proofreader:
Anna Bendiy
Academic Editors:
Natalia Shlikhta
Tetiana Vodotyka
Historical Advisor:
Yehor Brailian
Photographers:
Ihor Borysov
Oleksandr Strelnikov
Vitalii Sidorenko
Volodymyr Falin,
Semen Grozyan
Photo editor:
Maksym Bilousov
Publisher:
ist publishing

INNOVATION IN ISOLATION

The Story of Ukrainian IT from the 1940s to the Present

contents

INTRODUCTION	008

Part 1: 1947–1991

chapter 1
COMPUTER SCIENCE TRAILBLAZERS	018

chapter 2
INTERREGNUM	032

chapter 3
REBOOT	042

chapter 4
THE HEYDAY OF CYBERNETICS IN THE USSR	058

chapter 5
OGAS: THE SOVIET INTERNET	084

chapter 6
THE USSR AND THE RISE OF PERSONAL COMPUTERS	096

chapter 7
UKRAINE AT PLAY	104

chapter 8
AN ALL-UKRAINIAN PC	112

chapter 9
ZX MANIA	118

contents

chapter 10

THE LIFE AND DEATH OF SOVIET PCs — 124

chapter 11

THE USSR'S FINAL CHAPTER — 134

chapter 12

1990s: AFTER THE DOLDRUMS — 140

Part 2: Interviews with Modern Ukrainian IT Companies

MACPAW: EMPOWERING HUMANS WITH TECHNOLOGY	150
READDLE: "THANK YOU FOR TRUSTING US 200M TIMES!"	162
GRAMMARLY: IMPROVING LIVES BY IMPROVING COMMUNICATION	170
DEPOSITPHOTOS: SLOW AND STEADY DOESN'T WIN THE RACE	178
PETCUBE: KEEPING YOUR PET HAPPY WHEN YOU'RE AWAY	184
AJAX SYSTEMS: AN IDEA IS NOTHING, EXECUTION IS EVERYTHING	190
REFACE: ONE SELFIE IS ALL YOU NEED	200
JOOBLE: MATCHING TALENT AND COMPANIES SINCE 2006	208
BIBLIOGRAPHY	218
ACKNOWLEDGEMENTS	222

Engineers at work on SM-4, a Soviet PDP-11 compatible minicomputer produced at a range of facilities in Ukraine in the 1980s

Introduction

The story of any scientific discovery or achievement is, first of all, a story about the people behind them, their times, and the political dramas playing out in the background. This book's main focus is 1947 through 1991, when Ukraine was part of the USSR as its constituent republic. A smaller share took place after Ukraine gained independence on August 24, 1991. This is why our book is divided into two unequal parts.

The first, larger, part reconstructs Soviet Ukraine's role in the development of computer science alongside its social and political reality. We present the key developments in IT over the span of several decades, from the years of scarcity following World War II up through the 1990s. The second, smaller, part showcases current Ukrainian tech startups active on the global market. The book's two parts are divided by the hardscrabble 1990s, the tumultuous years when newly independent Ukraine was trying to establish itself from the ruins of the former Soviet Union.

Two things motivated us to write this book. Few outside of the post-Soviet space recognize the names of Sergey Lebedev or Victor Glushkov, two geniuses of cybernetics who brought the other side of the Iron Curtain into the computer age. Yet, it is thanks to their work that other landmark scientific achievements of that time were possible. Our book is an attempt to restore justice: to tell the story of the scientists who strove to make life in the USSR a bit better, in defiance of numerous material, technical, ideological, and political barriers.

Secondly, we wanted to present this dynamic chapter in the history of Ukraine, which was the beating heart of cybernetics in the USSR. Apart from Britain, no other European country (and certainly no other Soviet republic) had developed computing technology until scientists built the MESM (or "small electronic computing machine") in Kyiv, in 1950. Although the USSR is no longer remembered for its contributions to cybernetics, it was very nearly on par with the United States in the early decades of computing. Ukraine during this period could be fairly called the center of computer innovation among the Soviet republics.

Ukrainian cybernetics history can be broken down into several distinctive periods. The first is the postwar period, beginning after German forces were repelled from the USSR through the 1950s. On one level, this period was driven by passionate, self-sacrificing scientists who pushed through postwar scarcity and ideological strictures to invent technology that would, quite literally, change the world. The underlying narrative is, of course, one of ideological confrontation between the victorious Soviet empire and its rivals to the west. These years saw no less an achievement than the invention of the first electronic computing machine in continental Europe, and it happened not in a state-of-the-art laboratory but in a bombed building on the outskirts of Kyiv.

The 1960s saw the fruit of these scientists' work realized. Nikita Khrushchev's unprecedented

Reviewing the results of calculations made by the first specialized digital electronic machine in the USSR and continental Europe

↑
Engineers at work on ES EVM-1045, a unified system of mutually compatible computers, one of the Soviet analogues to the IBM system/360

denunciation of Stalin's "Reign of Terror" and the comparative openness that followed are to thank for that. Soviet ideology, which spurned religion, believed in the omnipotence of science. It was a golden age of scientific exploration, including in cybernetics. Computers powered the major developments in Soviet space exploration, nuclear technology, and the defense sector that raised the USSR to the status of a global superpower. It was in this period that OGAS — a bold prototype of the internet for Soviet production systems — was born. Unfortunately, the decade was a momentary rise before a fall that few could have predicted at the time. After the death of Victor Glushkov, nicknamed "the father of Soviet IT," the USSR never went back to developing these types of independent, unique projects.

The freedom scientists experienced during the brief "Thaw" was followed by what is now remembered as the "era of stagnation" — the 1970s — and the *perestroika* of the 1980s, ending in the Soviet Union's collapse in 1991. After the microprocessor hit the market in 1971, original Soviet projects were eventually scrapped and factories churned out clones of Western computers like IBM, instead. This can be read as a result of increasing skepticism towards traditional communism — evidently, the Party's promises of the October Revolution failed to materialize — in favor of "developed socialism." In style, entertainment, and behaviors like smoking and drinking (long discouraged by Soviet propaganda), Soviet people took cues from the West.

In computing, the American IBM was considered the gold standard. The proliferation of clones is evidence of the pervasive attitude in the Soviet Union, with regards to technology and other spheres, that "our people just can't do any better" — and no one expected them to. In fact, Soviet cybernetics had already reached its tipping point in the late 1960s when the government decided to follow the path of least resistance and clone Western technology in lieu of focusing on homegrown innovations. Why this decision was made at a time when the prospects for Soviet computers were spectacular remains, even today, a mystery. In this respect, researchers have more questions than answers.

It's worth mentioning, too, that there were some serious missteps in the area of software development. The Soviet Union made computer hardware but paid little attention to programs. State enterprises would typically receive boxes with "gear" and then do their best to assemble the machines and work out bugs on their own. Software options were available, increasingly in the age of the microprocessor, but they couldn't compare with the vast library offered by American and Western European companies. It would have taken millions of rubles and years of hard work to develop an operating system, let alone application software. Perhaps this was another consideration behind the decision to turn to clones.

In the 1970s, the USSR quickly backslid from its relatively high position in the 1950s and 1960s and became fully dependent on what could be copied from the West. The CoCom embargo, introduced at that very moment, only fueled the backpedaling of Soviet designs. As a result, the USSR was doomed to trail behind forever (several years could pass between the release of an all-new product in the West and the appearance of its Soviet clone) and make do with weak processors, as it failed to create technological capabilities (or resources, at that) to mass produce hardware components newer than those of the late 1970s.

The transition in the late 1970s to microprocessor technology (and the consumer computing industry) further debilitated any attempt at a lasting Soviet computing industry. Technologies like that can flourish only in market economies in which consumers have sufficient disposable income. The Soviet Union was not that place. Despite government reports showing a decent standard of living, the reality was that many people couldn't afford as much as a refrigerator, much less a personal computer. In Moscow, the "thriving" Soviet capital, bags of food hanging from balconies were ubiquitous in winter.

The lack of reliable telephone communications was a harder problem to solve. Technologies evolve in relation to the size of the potential customer base. At the same time, the only financial backer of computing projects was the state, and the state cared only for those industries that were directly related to the military industry. By the early 1980s, the arcade industry in the United States generated $8 billion USD a year, more than profits from music and cinema combined. In the Soviet Union, however, the only good investment in consumer technology was the cheapest one: pirating from the West.

So long as technological strength depended on large computers and heavy industry, the army, or the energy sector, the USSR could keep abreast with the United States. The Soviet Union had considerably fewer computers, but they kept pace with their Western counterparts. But as soon as the world moved to microprocessors, the USSR was left in the cold. After two decades of Soviet achievements in cybernetics, the incentive for Soviet computer design simply evaporated. By the 1980s, the USSR was already unable to compete. This is why a USSR of the 2000s is simply impossible to imagine. The reality is that the Soviet Union could not have built itself into a global power without computers — and could not sustain itself as one with them. Once computers left the laboratory and entered mainstream culture, the Soviet Union's fate was decided.

The consumer technology that was accessible for Soviet citizens came in the form of programmable calculators, arcade games, and DIY computer instructions printed in hobbyist magazines. In the 1980s, personal computing filled the same role as stamp collecting or film photography: it was an escape from reality. Personal computers assembled at home by a small number of tech enthusiasts offered a passageway into a parallel world. Ukraine under Soviet rule was already a country of dreamer engineers. The multimillion print run of *Radio*, a popular tech magazine, speaks for itself.

Yet to paint the personal computing trend as simple escapism ignores an omnipresent reality of life as a Soviet citizen. The state could not, or did not, provide for domestic needs. If you needed something, you had better find a way to get it on your own. The call from the state to "do it yourself" was more about necessity than ideology. Yet the grassroots computing movement petered out by the mid-1990s when the collapse of the Soviet Union cued the entrance of cheaper materials, and affordable IBM PC clones hit the market.

At the same time, the Soviet government, which had been at the industry's helm, was in pieces, barely capable of maintaining authority in its remaining territory. After the Soviet Union dissolved and the carefully planned economies in its former republics collapsed, computing research and development were simply out of the question. This cleared the way for a third wave in Soviet IT: entrepreneurship. Those who had an interest bought computer parts and assembled them for their friends. Soon enough, the first development companies began appearing. With cheaper materials available and government restrictions gone, the demand for computers grew with each passing day.

The last great shift came not with the fall of the Iron Curtain, but when borders truly disappeared: the invention of the internet. It was a period when those early software engineering companies started orienting themselves to the international market. That international focus remains true of Ukraine today. Ukraine is rapidly building a reputation as Europe's Silicon Valley, both because of the abundant outsourcing talent and the strong Ukrainian-made tech products and startups on the global market.

A NOTE ON RESEARCH

Our approach to writing this book was to go beyond the official histories guarded in archives. Beyond primary and secondary print sources, the story of Ukrainian tech that we have reconstructed here comes from dozens of interviews with those who personally knew the key players. Among them are the colleagues, relatives, and students of the cybernetics pioneers Victor Glushkov and Sergey Lebedev. In several cases, we were lucky enough to interview the scientists themselves, among them the creator of the Dnepr computer, Dr. Borys Malynovskyi, just before his death in 2019. Sadly, he was not able to see this book in print.

Authors

Volodymyr Nevzorov
Vladimir is a developer, author, and owner of the publishing house Sky Horse. Sky Horse has published more than forty books on technology and Ukrainian tourism, including *Apple: The Evolution of a Computer* and *Chernobyl Zone Through the Eyes of* Stalker Volodymyr himself writes widely about the history of computing in post-Soviet countries as a blogger and the author of a recent encyclopedia of Soviet gaming. Born in Moscow, he has been living in his father's native Ukraine since 2005.

Victoria Ugryumova
Victoria is an essayist on topics that include Kyiv history, medicine, and popular science. She served as Editor-in-Chief of the medical publishing house Professor Preobrazhensky between 2007 and 2010, and is the author of twenty-four works of popular fiction. Apart from co-authoring and editing this book, she recently completed an extensive photographic encyclopedia, *Postage Stamps of the USSR*.

Editor

Rachel McVey
A marketer and Fulbright alum, Rachel is passionate about telling the stories that matter. Her work includes projects in the promising Eastern European startup space, as well as collaboration on cultural initiatives for the US-Hungarian Fulbright Commission and Kyiv's Mystetskyi Arsenal National Art and Culture Museum. Rachel holds degrees in history and sociology from the University of Pittsburgh.

A thank you from the authors

We, the authors, would like to express our sincere gratitude to MacPaw for this opportunity to publish a book that preserves and passes on an important chapter in our country's history. The story of computing in the Soviet Union is, in many respects, a Ukrainian story. Ukraine today is known widely for its resilience and resourcefulness while defending itself against Russia's invasion. For us and, we hope, for you as readers, the story of Ukraine's role in cybernetics reveals a legacy of determination and progress in the face of many great obstacles.

↑
An employee of the Institute of Cybernetics in Kyiv holds the EC 5053 Magnetic Disk Pack. It consists of six disks of a total capacity of 7.25 MB

For the bulk of the 20th century, Ukraine had been ruled by Moscow and the Communist Party.

This is the story of how the nation transformed itself from a former Soviet republic into a budding innovation hub of Europe — and it starts all the way back in 1947.

chapter 1

↑
Small Electronic Calculating Machine (abbreviated "MESM" in Russian), late 1940s

In the Soviet Union, anything and everything pertaining to the creation of an electronic computer was top secret business. Special authorization was required just to take pictures of these machines, which accounts for why there are only a few surviving photos of the USSR's first computer.

Computer Science Trailblazers

For the close to fifty years of the Cold War, the prospect of nuclear warfare between the Soviet Union and the United States threatened to bring the world to sudden destruction. But the technology that once pushed the world to the brink of nuclear disaster was only possible because of a DIY computer project started in the lobby of an abandoned Kyiv rest house in 1947 — the electronic computer.

Sergey Lebedev, the director of the Energy Institute at the Academy of Sciences of Ukraine, led the team of engineers that would eventually invent the first computer prototype built behind the Iron Curtain. As the world entered the computer age, their invention, hand-in-hand with the work of scientists in other Soviet republics, laid the groundwork for the Soviet Union to become a global power. Yet the significance of their achievement would not be recognized for fifty years.

Until the 1940s, the Soviet Union, like the rest of the world, relied on human calculations and simple mechanical computing machines for engineering and defense. In the late 1940s, not even an army of engineers equipped with arithmometers could have performed the calculations needed for nuclear modeling or predicting ballistic missile trajectories quickly enough.

In the late 1940s, scientists in the USSR were tasked with the job of creating an electronic computer that would keep Soviet defense on par with the United States, where cybernetics was just beginning to be explored. But there was a catch: they had to replicate the machines Western scientists were making, without knowing what, exactly, those scientists had made. Just for reading a Western journal, Soviet scientists could be charged with espionage and sentenced to 10–15 years in a Gulag.[1] Only a few select experts were allowed to apply for the reading permit required to read foreign journals at restricted-access research libraries.

The Soviet government in the late 1940s and 1950s had a love-hate relationship with technological innovation. It was as clear to party leaders as to scientists and military generals that technology was the only way to remain a global superpower. But it was a special quirk of the USSR that ideological considerations often outweighed pragmatic ones. As Cold War tensions grew, the Soviet Union launched an all-out war against all things Western. Western fashion, art, and science were all taboo. So was electronic computing.

For many years, the Party line was that computer science, or cybernetics, was a "bourgeois pseudoscience." In their quest to uphold "true communism," Soviet leaders essentially asked their scientists to build a computer with their hands tied behind their backs.

Yet in spite of these roadblocks, a handful of determined scientists accepted the government's call to develop the first Soviet computer. Sergey Lebedev is the reason they succeeded.

↑
Sergey Lebedev (1902–1974)
Soviet scientist, member of the USSR Academy of Sciences, a pioneer in cybernetics, c. 1930

1/ Borrowed by Joseph Stalin from the Russian Empire, the brutal Gulag system of forced labor camps was used to eliminate "class enemies" as well as "nationalists" — a term applied to cultural figures in republics colonized by the Soviet Union. Kateryna Shkabara and Kateryna Yushchenko, key inventors in this period, were initially thrown out of school after their parents were repressed on suspicion of holding pro-Ukrainian views.

Ukraine in the postwar period was the hub of scientific innovation in the Soviet Union. In 1946, the Academy of Sciences of Ukraine had 15 researcher positions, but only one of them was dedicated to the development of an electronic computer. Lebedev was an easy choice for the job. He had distinguished himself as a defense engineer in Moscow, where he began learning about cybernetics from foreign journals.

He did not fit the mold of the scatter-brained, eccentric scientist that often appears in books and movies. The Lebedev family socialized with the Soviet elite, frequently hosting famous writers, athletes, and actors at their home for parties and concerts. Lebedev himself was outgoing and sharp-witted. His knack for making friends put his name on the radar of the President of the Academy, and Lebedev was appointed to both the Academy and the Electrical and Thermal Engineering Institute in Kyiv, Ground Zero of modern computing in continental Europe.

Like most scientists in the USSR, Lebedev worked for passion rather than money. For an extended period of time, Lebedev and his wife lived in a cramped apartment where the kitchen doubled as his office. He was not used to working alone in silence, and when he was given a private office in Kyiv, he continued working at the dinner table, even when he and his wife were entertaining. Lebedev often made notes on scraps of paper or on cigarette packs, keeping up the dinner conversation and mulling over a new design at the same time.

In 1947, Laboratory #1 for scientific modeling and computing was set up in Kyiv at the Ukrainian Socialist Republic's Institute of Electrical Engineering at the Academy of Sciences. Its sole task was to create and launch an electronic computer model (ECM) as quickly as possible. Lebedev reported to a supervising committee that other countries had taken between five and ten years to develop a working ECM. His goal was to do it in two.

The project was a long shot from the start. Soviet scientists had little to work with, except for the knowledge that such a computer had already been built in Britain and the USA. For all their engineering expertise, neither Lebedev nor his colleagues in Laboratory #1 had any experience building computers. At first, the lab was home to no more than ten researchers, but the staff doubled within a year. Most of them were graduates fresh out of university who had returned to their studies after the war, some of them wounded or on crutches.

Then, there was the problem of space. The planned machine would be enormous, requiring 100–150 sq. meters (1,000–1,600 sq. ft.) of floor room. The auxiliaries — generators, automated controls, and energy accumulators — would need the same amount. And it could not be

Kyiv Polytechnic Institute at the start of the 20th century. The electric tram pictured was among the first of its kind in Eastern Europe, installed in 1892.

The period from the late 19th century until the outbreak of World War I was quite productive for Ukraine in many respects: industry was thriving, people were making fortunes, and cities were rapidly growing. However, Ukraine remained dependent on the interests and decisions at the respective political centers of the Austro-Hungarian and Russian empires.

As its name suggests, Kyiv Polytechnic Institute (KPI) was founded in 1898 in Kyiv, a forward-looking Ukrainian city ruled by the Russian Empire since the mid-17th century. Kyiv's governors dipped into their own pockets to develop it. Industrialists and merchants did their best to keep up with them.

They sought to leverage state-of-the-art technology, viewing the American lifestyle as a benchmark of sorts, while still trying to preserve European traditions. This mix of old aristocracy and cutting-edge technology shaped the unique culture of Kyiv that still distinguishes it from other Ukrainian cities today.

Ukrainian scientific institutions were known for producing (and attracting) some of the best scientists in the region. Back in 1875, Mykhailo Avenarius, founder of the Kyiv school of experimental physics, established an experimental physics laboratory, the first one of its kind in Ukraine. He determined the critical temperatures for liquids and many other substances that were later recognized as fundamental physical measurements.

In the 20th century, dozens of notable scientists would graduate from the university, among them Serhiy Korolev, founder of the Soviet Union's space program, and Igor Sikorsky, the world-famous helicopter inventor nicknamed "Mr. Helicopter," were among KPI's notable graduates.

← Khreshchatyk Street, Kyiv's main boulevard, in the early 1950s

built just anywhere, either. All the workshops and service units involved in building such a powerful machine also needed to be easily accessible.

This kind of location simply didn't exist at the stately Academy of Sciences. So Lebedev and his team set up shop a few kilometers south of the Institute, in the nearest building large enough to accommodate such a massive tenant — a bombed-out inn for religious pilgrims in Feofaniia, the former grounds of a cathedral in the suburbs of Kyiv.

Surrounded by idyllic lawns and lush forest, the would-be laboratory was in horrendous condition. In the machine room, the temperature reached as high as 40°C (105°F) in the winter, even with the windows open. In the summer heat, the tubes became so hot that the machine often malfunctioned. But this was a Kyiv recovering from three years of Nazi air raids during World War II. Feofaniia was only one of many neighborhoods mutilated by bombing. Khreshchatyk, the main street of the capital city, lay in shambles. With the entire city disfigured by war, a building with all its walls intact was considered more than suitable for a lab.

Poor working conditions in the lab were matched by a punishing workload. The lab staff toiled away, putting in as many as 14 hours a day. They even worked on Sundays, the only day off Soviet citizens got during the post-war recovery period. The researchers often had to put in even more hours, working at breakneck speed to ensure that the machine was fueled, cooled, and monitored continuously.

Each day, a clanky shuttle bus would take employees to their makeshift lab. When it rained or snowed, the passengers would have to jump into the slush and push the bus to higher ground. Often, the bus would break down entirely and they would make the rest of the trek on foot, following the old pilgrims' path to Feofaniia.

Sergey Lebedev in his study

↑
Most likely a fan from the MESM

2/ When revisiting this moment in history from the vantage point of post-February 2022, it is impossible not to draw comparisons with present-day Ukrainians, who have lost their lives to Russian land mines in the course of their daily work.

Many on the team sidestepped these transportation difficulties by taking up residence in the lab. Ignoring the lack of amenities, they built a makeshift dorm on the second floor, right above their invention, which they had affectionately nicknamed "Kiddo."

There was no running water, and outhouses stood on the edge of the forest, wash basins beside them. In the winter, they bathed with icy water from the nearby lakes. In the summer, the machine's 6,000 vacuum tubes made the place as hot as a desert. But incredibly, the scientists and technicians behind the ECM pushed through two years in these conditions. Lebedev himself lived at the lab all year round. His wife and kids joined him for the summer.

Sharing one space and a multitude of everyday problems month after month, it's no wonder that lab employees quickly became like family. In their little time off from work, they made a clearing in the woods and set up a volleyball court. Dozens of memoirs about that postwar period make a unanimous claim — despite living hand-to-mouth, the Laboratory #1 engineers always kept their chins up.

At the same time, the recent war continued to visit danger on the team. The forest was still littered with unexploded munitions and mines that claimed the lives of several lab employees during these years[2].

Still, there were some perks to the unusual residence. The most obvious was something very basic: access to nutritious food. Ukraine in 1947 was suffering from hunger. Even though, during this period, Soviet authorities were exporting Ukrainian grain abroad, building Ukraine's reputation in the West as a land of fertile, food-giving fields. But behind the glitz of Soviet propaganda was a shocking reality: this grain was taken from Ukrainians by force, leaving entire villages to starve.

In Ukrainian cities, bread, sugar, flour, and other pantry staples were strictly rationed. In Feofaniia, on the outskirts of Kyiv, the single grocery store offered its shoppers only a minimum assortment of canned goods and 777, a cheap red wine palatable to only the most devoted drinkers. Food products could be found at markets for exorbitant rates, but on researchers' salaries, the Laboratory #1 scientists could not afford much.

Living at Feofaniia gave the team the chance to supplement their meager Soviet rations with food from the surrounding forest. Berries, mushrooms, and game found its way onto their plates, along with vegetables occasionally raided from the garden of the Academy of Sciences president Oleksandr Palladin, whose summer cottage was in Feofaniia.

On top of the dire working conditions and the overall postwar slump, the researchers had to deal with insufficient funding. It wasn't the most favorable time to start working on the computer, and the government's ambivalent attitude towards cybernetics certainly had an effect on the project. The lab had very limited resources and lacked even the most basic parts for its equipment.

As the lab director, it fell to Lebedev to juggle the responsibilities of scientist, administrator, fundraiser, and logistical support. Dealing with the issues of poor supply, the Soviet scientists had to think outside the box. There was always a scarcity of high-quality vacuum tubes, so the lab used nonindustrial tubes for radio sets instead. This solution was far from perfect, however. Roughly a dozen of them would burn out every hour. Eventually, the research team developed a special testing system that drained the tubes considerably less, so the machine could run up to seven hours continuously.

Other parts they bought from depots of captured enemy equipment, where spare parts could be bought for peanuts. Most of the items stored at the depots were outdated prewar models and the gears, pipes, and scrap metal the team acquired from them were a far cry from the state-of-the-art equipment that the American engineers were using to build their own ECM.

When it came to funding, the lab earned a large share of its income by producing equipment for the Soviet military. For example, they designed a system that allowed pilots and rocket engineers to run tests on models, rather than on real equipment. The military paid generously for these developments, and the lab, in its turn, used this income to finance its ECM project.

Thanks to the team members' ingenuity and the cash inflow from the military, the project made rapid progress. The command switch unit occupied the most conspicuous spot at the lab. It was a 60×60 cm (23×23 in.) square with cheap black tube diodes. By the end of 1949, "Kiddo" was given an official name, a Small Electronic Calculating Machine. Now, in the middle of it there were four glowing letters, symbolizing the Russian abbreviation: MESM.

The first problem that the MESM had to solve was comically simple: what two times two equals. But it took a while for the machine to give a correct answer. The first results were discouraging. When prompted to solve 2×2, it answered five, nine, a hundred — anything but four. It was especially hard in the summer heat, because the machine could work only at night, when temperatures were lower. The researchers had to work out the kinks through trial and error, as they didn't have any theoretical data and had to fine-tune the system from scratch. Developing an electronic computer was half science, half art.

The two-times-two stage didn't last for long, though. In the fall of 1951, the machine was given a real-life test in ballistics, and it identified, for the first time, an error made by humans — and not just any humans, but highly-qualified professionals.

Initially, the MESM designers assumed that the machine had failed. Sergey Lebedev alone believed that it was right. While the lab staff checked and double-checked the machine, tuning it up and re-adjusting its settings, he spent almost the whole day verifying his colleagues' calculations to the ninth digit.

As the story goes, the next morning, he came in smiling (an unusual occurrence for him at that time), his glasses, resting on his forehead, a signal that he'd succeeded. "Stop tormenting the machine — it's right," he said. "Humans are wrong." It turned out that, in an incredible coincidence, two experts had made an error in the same spot — and the machine promptly identified it. As the team achieved success after success, the MESM project was made top secret. In all the documents, the first Soviet electronic computer was listed simply as Project #1.

Then, in late 1951, Lebedev got the news he and his team had been hoping for since starting their work four years earlier: a government commission from Moscow would be arriving to review the MESM and accept it into service.

But even after all that, the scientists' path was not clear. On November 30, 1951, Project #1 went up in flames. The vertical wiring harness malfunctioned and three units were partially damaged in the blaze. There was no time to spare. Working around the clock, the team managed to repair the damage and present "Kiddo" in working order to the reviewers.

The government quality control test lasted a nerve racking three days. The commission members set tasks for the MESM one by one. The machine's inventors looked on as anxiously as new parents. A smart "Kiddo," the MESM passed the test with flying colors, and on December 25, 1951, the government commission accepted the MESM, the first electronic computer in continental Europe, into service.

The team's success was and remains a testament to their perseverance — and to their passion for cybernetics. Later, when the MESM was handling seven-digit numbers, seasoned scientists who visited the lab to see the brand new "marvel of engineering" would ask the machine the same question over and over: what's two times two? They shook their heads in wonder when the MESM answered correctly.

Sergey Lebedev with his wife and daughters ↓

Khreshchatyk, the main street of Kyiv, in 1944. The first creators of Soviet computers worked in such conditions.
↓

MESM, late 1940s

The building housing Laboratory #1 had been transformed more than once to suit the needs of the day. First built as an inn for pilgrims visiting St. Panteleimon Monastery near Kyiv, it had later been converted into a mental institution by the vehemently atheist Soviet government.

The MESM occupied 60 sq.m. (645 sq. ft.) — an entire wing of the aged building — and consumed a whopping 25 kW of electricity. The processor with a clock rate of 5 kHz performed about 3,000 operations per second, solving a variety of tasks commissioned by a number of government institutions, mostly those related to the military.

The MESM started its actual service in January of 1952 and finished it in 1957. It was later donated to the Kyiv Polytechnic Institute for research purposes and was dismantled two years after that. Only some fragments of this giant, including (presumably) one of the cooling fans shown in the following photo, have survived until today.

↑
The Felix arithmometer

The Felix was the standard Soviet arithmometer. From the late 1920s to the late 1970s, a dozen factories produced several million of these mechanical calculating machines, used mostly for long multiplication and division operations.

A heavy (4–6 kg, depending on the model), yet quite small desktop "calculator" was an affordable device. After the Soviet currency revaluation in 1961, its price was just 11–15 rubles. (The average monthly salary at the same time was between 60 and 70 rubles, or $60 to $70 according to the official exchange rate of the day.) The forerunner to the Felix was an arithmometer designed by Willgodt Theophil Odhner, a Swedish engineer and entrepreneur who worked in the Russian Empire. From 1890, it was mass-produced in Saint Petersburg.

After the Bolshevik revolution of 1917, the Swedish inventor's factory (his heirs' property, by that time) was nationalized, and the Odhner's arithmometer was renamed *Felix* in honor of Felix Dzerzhinsky, the Soviet Bolshevik revolutionary.

↑
Norbert Wiener (1894–1964), American mathematician, philosopher, and originator of cybernetics and artificial intelligence theory

Cybernetics (from the ancient Greek κυβερνητική, "good at steering") is a science dealing with the general principles of information collection, storage, transformation, and transfer in complex systems of control. The systems of control pertain here not only to technical systems, but also to any other ones, such as biological, administrative, or social.

French physicist André-Marie Ampère used the term "cybernetics" in 1834 for the first time since the ancient Greeks in his classification of the sciences to designate the still nonexistent science of government. But the term was soon forgotten. It was Norbert Wiener who revived it, titling his book *Cybernetics: Or Control and Communication in the Animal and the Machine*, published in 1948. The date of Wiener's publication is generally considered to be the birth of cybernetics as an independent science. Wiener defined cybernetics as a science "about control and communication in the animal and the machine." He did not mention human society in his definition.

In 1954, though, Wiener published another book, *The Human Use of Human Beings: Cybernetics and Society.* However, both books were mostly theoretical and did not offer any practical methods for using the brand-new science. Around the same time in the Soviet Union, Victor Glushkov outlined his pet project-computerized industrial control systems. Industrial control systems put computers to use for state organizations and were already at work in the United States. Glushkov argued for various practical applications in a number of spheres ranging from the automated processing of scientific data to the management of complex economic systems.

"THE MOST ACTIVE AND POLITICALLY MOST CONSCIOUS CITIZENS IN THE RANKS OF THE WORKING CLASS AND OTHER SECTIONS OF THE WORKING PEOPLE UNITE IN THE COMMUNIST PARTY OF THE SOVIET UNION (BOLSHEVIKS), WHICH IS THE VANGUARD OF THE WORKING PEOPLE IN THEIR STRUGGLE TO STRENGTHEN AND DEVELOP THE SOCIALIST SYSTEM AND IS THE LEADING CORE OF ALL ORGANIZATIONS OF THE WORKING PEOPLE, BOTH PUBLIC AND STATE."

— ARTICLE 126 OF THE USSR CONSTITUTION, DECEMBER 1936

THE GODLIKE ROLE OF THE PARTY UNDERLAY EVERY EPISODE IN THE SAGA OF SOVIET CYBERNETICS (AND THE LIVES OF ITS PROTAGONISTS).

ALTHOUGH UKRAINE, LIKE EACH SOVIET REPUBLIC, WAS OFFICIALLY A CONSTITUENT OF THE SOVIET UNION WITH ITS OWN SUPREME COUNCIL, IN PRACTICE, ALL POLICY DECISIONS WERE MADE IN MOSCOW.

chapter 2

↑
Sergey Lebedev and his colleagues at work in
the Automation and Energy Laboratory in Kyiv

Interregnum

In the closing days of 1951, Sergey Lebedev and his team of researchers overcame incredible odds to create the first electronic computer in mainland Europe. For one week, they rubbed shoulders with top officials and showed off the power of their machine. They were even shortlisted for the Stalin Prize — the most prestigious award that a scientist in the USSR could hope to receive at that time.

After years of laboring night and day in a makeshift lab, struggling to magic scavenged parts into a working computer, it looked like the scientists would finally get the recognition they deserved.

And then, nothing. Unexpectedly, the award committee of the Academy of Sciences refused to sign off on the accolade. The official reason for the snub was that the nomination had been submitted to the Committee too late. Some people blamed it all on Lebedev's ill-wishers, claiming they plotted against him and his team. It could have been simply a symptom of the Soviet government's incoordination, resulting from a system that rewarded loyalty over competence. Undeterred, Lebedev prepared a report on his team's performance and presented it to the Academy of Sciences' Presidium behind closed doors.

A week later, the core team received an answer. They would be recognized with only a commendation. The years they devoted to bringing the USSR into the computer age earned them nothing more than a mark in their employment record books. Whatever really happened, one thing could not be clearer: the Academy of Sciences did not consider the Laboratory #1 project a top priority.

→
Soviet Science announces the winner of the Stalin prize on March 18, 1950

Text in translation: "To Soviet Science. Congratulations to our laureates of the Stalin Prize on receiving the honorary title. May they have continued success for the benefit of our Homeland."

"An Outstanding Invention and Its Creators: L. V. Tsoukernik; Our Sergey Alekseevich."

In January 1951— almost a year before the triumphant launch of the electronic computer — the academic board of the Academy of Sciences' Electrical and Thermal Engineering Institute convened behind closed doors to harshly criticize the project. The board took Lebedev to task for the smallest details, betraying their ignorance of computer science. "The term 'logical operations' should not be applied to a machine, for it cannot perform logical operations. Some other term should be used," one member fussed. Clearly, the board and the computer scientists were not on the same page.

Lebedev's team was operating in a heavily ambivalent environment: computing was a somewhat necessary evil, and it certainly wasn't the scientists themselves who determined what constituted "necessary." Even as Lebedev and his team came into the final stretch of the project, the government opinion on cybernetics flip-flopped. In the 1950s, cybernetics was widely condemned for its "cosmopolitan" components. But at the same time, the Soviet Union needed computing machines to grow its military capabilities.

At the helm of the Academy was Oleksandr Palladin, a renowned biochemist who didn't know a thing about computers. What he did know was how to respond correctly to Moscow's stance on cybernetics. In early 1951, Palladin suggested that Lebedev move the MESM — almost ready by that time — to Moscow. There, the government would be able to keep a closer eye on the project. Lebedev refused, saying that it might delay the project's completion by almost a year.

Completing the project in Moscow could have well given Lebedev the political connections he lacked in Kyiv, the kind of connections that would have earned the team the Stalin Prize instead of a check mark in their employment books. But Lebedev craved the status of invetor number-one more than a prize. His strategy — if he actually had one — paid off in the end.

As soon as the government commission accepted the MESM into service, Lebedev was offered relocation back to Moscow and an opportunity to manage another Laboratory #1, this time at the Academy of Sciences' Institute of Precision Mechanics and Computer Engineering of the Soviet Union. The Institute just started to develop a Large Electronic Computing Machine (BESM, an acronym for its name in Russian). Lebedev accepted the offer.

His team stayed in Ukraine. Their recollections vary. Some of his colleagues think that Lebedev could have easily brought them along to Moscow. Others believe that a committee within the Communist Party had called him in and "strongly recommended" that he leave his colleagues behind. There's no way to know for sure what really happened. In any case, the result was the same.

↓
Laboratory # 1 Kyiv, early 1950s. The Soviet system encouraged women to study at universities and pursue careers in the sciences. As a result, the MESM team included many women, including Kateryna Shkabara, who worked as Head of Development for the MESM controls before going on to head the laboratory of physiological cybernetics at the Institute of Sciences of the Ukrainian Socialist Republic.

Lebedev's career took off.

He had left Moscow second-best in his field only to return later as the undisputed leader, the creator of the first electronic computer in continental Europe. Lebedev gave up his position at the Academy of Sciences of Ukraine in favor of one higher in Moscow's hierarchy: Chair of the Institute of Precision Mechanics and Computer Engineering. He held the position for twenty years, supervising the development and launch of no less than eighteen computers. In this particular story, Lebedev was the undeniable winner — the Soviet Caesar, who fled Rome to Gaul in the face of political threats and then, having proven himself on the battlefield, returned to a hero's welcome.

Meanwhile in Kyiv, the prospects of Laboratory #1 and Ukrainian computer engineering in general were not as bright. The incoming director of the Institute of Electrical Engineering at the Academy of Sciences was stunningly dismissive of the department he was supposed to oversee.
"I don't need any computers!" was his constant rebuff to requests from the cybernetics team.

One episode captures the absurdity of Soviet bureaucracy in action particularly well. In 1954, the lab got a new boss — an energetic, freshly minted CSc (Candidate of Science, the Soviet

↑
A Laboratory #1 researcher at work, programming the first specialized digital computer SESM ("Specialized Electronic Calculating Machine" in Russian) in the USSR and Europe. It was designed to solve systems of linear equations with 400 unknowns each.

equivalent to a PhD), whose enthusiasm to take charge outmatched his competence in cybernetics.

After inspecting the MESM, the young director concluded that it was dusty and should be cleaned from top to bottom without delay. So, with a heavy backlog of projects growing weekly, the scientists stood back and waited while all of the MESM's six thousand vacuum tubes were purged of dust.

Unfortunately (but not surprisingly, for those who understood the machine), this heavy-handed intervention destroyed dozens of connections and brazed joints and brought Laboratory #1 to a complete halt for two full weeks. Still, the team stayed committed to developing cybernetics.

The scientists behind the MESM, a true breakthrough in Soviet technology, were some of the brightest minds in the USSR. These were people who were not content to simply sit back on their heels while bureaucracy warmed its engines. They took on the task of developing a new computer, called *Kiev* in Russian.

The team, minus Lebedev, built on the developments they'd already made to create a special-purpose computer — a vector processor, the first of its kind in Europe. They also did some research for the air defense forces (digital processing of radar data) and created a prototype of a digital onboard computer. However, these achievements were just a fraction of what the Laboratory #1 scientists were capable of. From 1951 to 1956, the lab was not working at full capacity.

Realizing the need for an intervention, several Laboratory #1 scientists who held membership in the Communist Party decided to go around the system. They sent an official letter to the Communist Party of Ukraine's (CPU) Central Committee this time, instead of to the Soviet Academy of Sciences. They didn't beat around the bush. "The way computing equipment is being handled (in Ukraine) verges on a crime against the state," they wrote. The researchers addressed their letter to the CPU headquarters for good reason. While the scientists were busy in their lab, outside their windows, the wheels of change were turning in Soviet politics. Stalin had died in 1953, prompting a reconsideration of the leader's harsh policies. There was a new attitude; hints of progress had started to appear. Moscow was changing — could its stance on computers change, as well?

A special-purpose electronic computer was built under Lebedev's supervision and launched in 1956. Zinoviy Rabinovich was the chief engineer on this project. The computer solved correlation tasks and systems of algebraic equations with up to 500 unknowns. In addition to its powerful capabilities, this machine had another distinctive feature. It utilized a vector processor — the first of its kind in Europe — that used ordered data sets, or "vectors" (in algebraic terms) as operands in specific commands. Unlike its vector counterpart, a regular, scalar processor could work with only one operand at a time. Vector processors performed specific tasks very well, so they were used in supercomputers for a long time.

↑
A special-purpose electronic computer — a vector processor, early 1950s

↑ The Kiev computer

THE KIEV COMPUTER

The team behind the MESM built another machine — the Kiev,* a special-purpose electronic computer. Boris Gnedenko, director of the Mathematics Institute of the Academy of Sciences, was the one who initiated the project.

The Kiev computer was one of the first machines in the world to use a high-level programming language, also known as *Address*. It was also one of the earliest systems for digital image processing and simple modeling of intellectual processes. The team built two computers — one was installed at the Academy of Sciences' Computation Center, and the other one was shipped to the Joint Institute for Nuclear Research in the town of Dubna, near Moscow.

They were led by Victor Glushkov, a famous mathematician and one of the founding fathers of the Soviet IT industry. They used this computer to experiment with the remote control of technological processes that took place 500 km (300 mi.) away at the Dniprodzerzhynsk Iron and Steel Factory. As early as the late 1950s and early 1960s, the team worked on artificial intelligence, trying to teach the computer to recognize geometric shapes, as well as uppercase and lowercase letters.

New software capabilities came with hardware innovations. The Kiev computer was based on vacuum tubes, just like the MESM, but, unlike that model, it had ferrite core memory with a storage capacity of 1,024 machine words that were 41 bits long. The machine also used a magnetic drum storage device with a total capacity of 9,000 words. Its average processor speed was 15,000 operations per second — an impressive rate at the time.

* The name of this computer and others appear in the Russian spelling, as Russian was the lingua franca of the Soviet empire.

Punch cards as a method of information storage and input appeared long before the onset of the computer era. In 1928, IBM introduced a popular 12-row, 80-column punch card format that was widely used in the USSR.

The simple IBM-card was just a piece of smooth stock paper measuring only 187.32 mm (7⅜ in.) long, 82.5 mm (3¼ in.) wide, and 0.175 mm (0.006 in.) thick. Quite unassuming, it had several perforated holes punched by a special machine that represented data-numbers, letters, or additional characters.

The IBM-card could store no more than 80 bytes. The whole process was quite time-consuming. It took a while to enter the information onto a punch card. A computer operator entered the data using a keyboard, then a keypunch made holes on the punch card at specific locations.

The operators often made errors, so the data entered on punch cards had to be double-checked by another device called a "verifier." A verifier operator had to key in exactly the same data, character by painful character, a second time. If the verifier identified that the second keying and the punched data were not the same, the punch card had to be discarded, and the user would have to have to start all over.

Then, the operator fed decks of punch cards into a computer card reader. The readers varied in speed (1,500–2,000 characters per second) – and could process 20-25 cards per minute. It's interesting to see how one technological standard shaped another one, as most computer text screens at that time displayed 80 characters per row – exactly the same number as in a punch card row.

Punch card operators worked in shifts. Equipped with one keypunch and one verifier, each of them could turn out no more than 200 punch cards per shift. This time- and labor-consuming process partially negated the positive effect of automation.

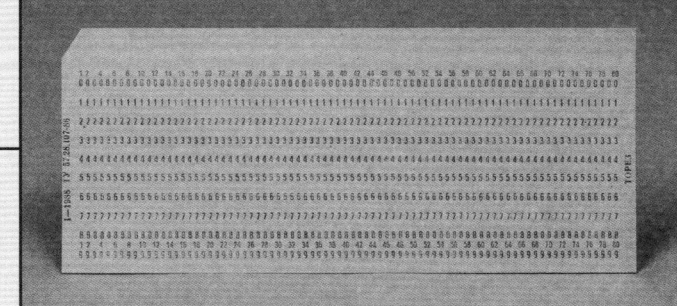

↑
Punch card

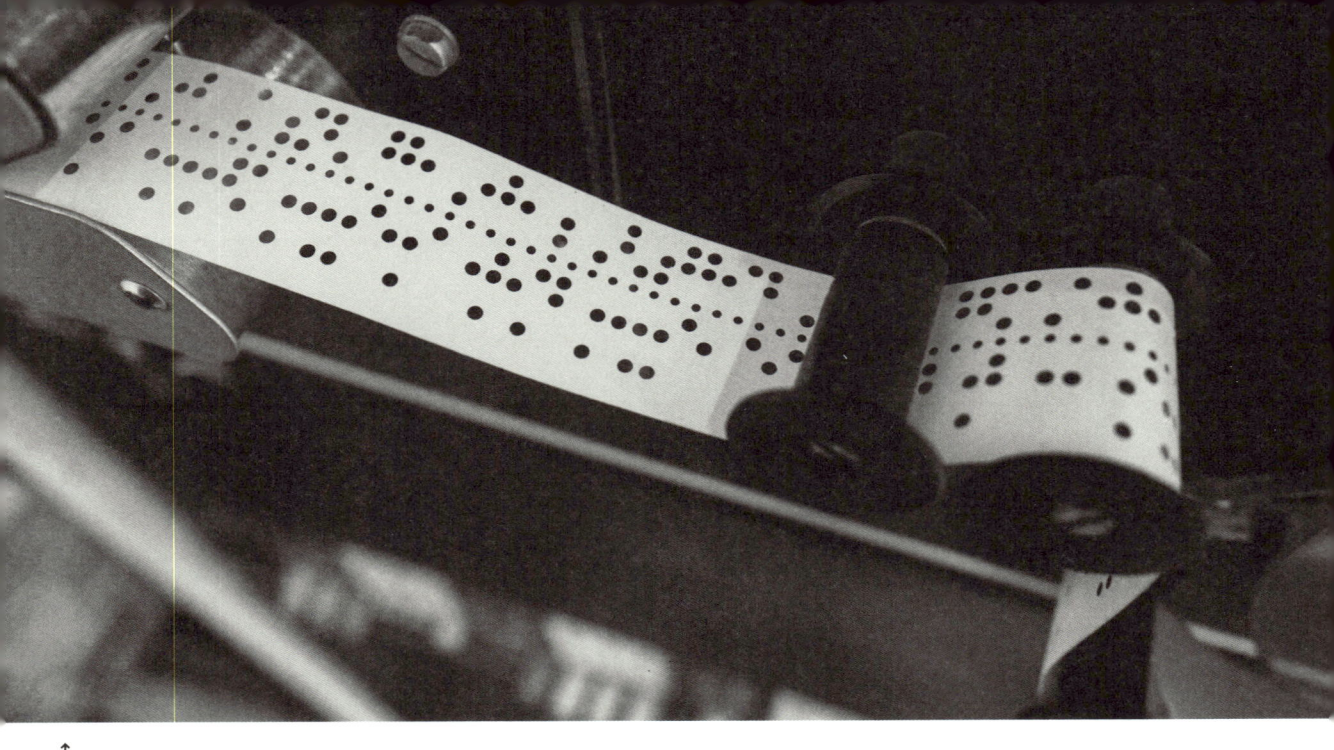

↑
Punched tape

← Kateryna Yushchenko, developer of one of the world's first high-level programming languages, called Address, and member of the USSR Academy of Sciences

→ Kateryna Shkabara is setting up the "Kiev" computer, Kyiv, 1954

Given the period, many readers may be imagining Laboratory #1 as a boy's club. In fact, two of Lebedev's key associates on the MESM project were women. Thanks to Kateryna Shkabara and Kateryna Yushchenko, some unique features of MESM were improved in the "Kiev" computer. Lebedev, for his part, learned from his Kyiv colleagues about the peculiarities of improving the unique capabilities of his MESM, and took them into account in the architecture of his later computers: M-20 and all computers in the BESM family.

Shkabara worked under Lebedev as the head of development and adjustment of the MESM controls and of the magnetic drum. Within ten years of finishing MESM, she was hired as the head of her own cybernetics lab at the Institute of Physiology of the Academy of Sciences of the Ukrainian SSR. There, she went on to play a leading role in the development of the Kiev computer, the next boundary-breaking computer behind the Iron Curtain. Her career would take her to merge innovations in technology and biology. In the 1960s, this would result in the first working model of a computer able to diagnose heart disease.

Yushchenko, for her part, would go on to develop the world's first high-level programming language, Address, in 1955. For seven years, Yushchenko held the position of senior researcher at the Kiev Institute of Mathematics of the Academy of Sciences of the Ukrainian SSR (1950-1957). The following year, she became the director of the Institute of Computing Technology of the Academy of Sciences of the Ukrainian SSR, where she formed an internationally-known scientific school of theoretical programming. Together with mathematician and scientist Vicktor Glushkov, she pushed so far as to create the foundation for artificial intelligence tools.

The proof of her brilliance was witnessed by the whole world in July 1975, as spellbound viewers watched the US' *Apollo* spacecraft dock with the Soviet *Soyuz* capsule mid-orbit. As the first international space mission was launched, and a new era in world history with it, it was Kateryna Yushchenko's Address language powering the controls.

The significant achievements of Shkabara and Yushchenko were all the more impressive in light of their personal lives. Like other Soviet-era scientists, they were not shielded from the difficult realities of life under an authoritarian regime. Ukrainian-born, both women suffered personally from the massive repressions Soviet authorities wielded against the Ukrainian population.

Shkabara's career prospects were all but shut down during her university studies when her parents — accused of "Ukrainian bourgeois nationalism" — were arrested as "enemies of the people" and sent to a Gulag in 1933. She was expelled from her program at the Kharkiv Electrotechnical Institute, along with her brother. Their determination saved both her career and, eventually, her mother.

Yushchenko's story is nearly the same. Her father, a history teacher, was arrested in 1937 for giving his children a tour of several Ukrainian Cossack (ancient warriors) heritage sites. Kateryna was just 17 years old at the time, full of ambition for her first year at the Physics and Mathematics Faculty of Kiev State Shevchenko University. As the daughter of an "enemy of the people," she was expelled from the university, nearly ending her career before it had even started. Yushchenko's mother's attempt to prove the father's innocence resulted in her arrest, as well: she and her husband were sentenced to 10 years in a Gulag. Only in 1954, after Stalin's death, Kateryna Yushchenko's parents were posthumously exonerated due to the absence of a crime.

chapter 3

↑
Victor Glushkov working at the library,
mid-to-late 1950s

Reboot

Although the scientists at Laboratory #1 didn't know it at the time, 1953 would come to be known as a turning point for cybernetics in the USSR: it was the year that Joseph Stalin died. Following a lengthy power struggle, Nikita Khrushchev emerged victorious in 1958 to lead the Soviet Union as the First Secretary of the Communist Party and the Chairman of the Council of Ministers. His bald denunciation of Stalin's crimes against the people ushered in a new era of openness, or at least, what passed for openness after Stalin's "Reign of Terror."

During the political and ideological "Thaw" that followed, most Gulag labor camps — millions of people had been imprisoned in the brutal forced-labor system since the 1920s — were shut down. Hundreds of thousands of prisoners became free, scores of talented scholars among them. Yet, it is worth noting that this was not a rosy epoch in Soviet history. While certain academic fields certainly benefited from Khrushchev's "openness," internment in Gulags remained the fate of political activists and cultural figures of non-Russian ethnicity.

Krushchev's first task at hand was to mend the Soviet Union's relations with the West. His agenda was to be the antithesis of Stalin — Stalin started the Cold War; Krushchev would bring about a "détente." But more importantly, he needed warmer relations with the U.S. in order to improve the economy of the USSR. The Cold War had become expensive; escalating the arms race further was economically unfeasible. The ideological competition with capitalism, however, only intensified.

Soon, Krushchev became the first Soviet leader to visit the United States. After more than a quarter-century of Stalin as the iron face of the Soviet Union, Khrushchev's earnest and sometimes colorful leadership took officials and citizens alike by surprise. At home and abroad, he cut a bizarre figure, threatening Richard Nixon with "Kuzma's mother"[1] at a diplomatic event and championing a disastrous attempt at an Iowa-style corn belt in the Soviet Union. "Catch up with and surpass America" was the anthem of schools, factories, and the laboratory in those years.

For the first time, cooperation in fields like agriculture and science became possible. It was under Khrushchev's leadership that the Soviet Union made the groundbreaking achievements in space exploration and nuclear science that would consolidate their status as both an enemy and an ally in innovation to the United States.

The many scientific breakthroughs of the "Thaw" era — achievements like the launch of the world's first space probe, the first human flight to space, and the space suit — would have been unthinkable a decade earlier. For the first time, Soviet scientists were allowed to travel abroad, participate in international conferences, and do research at American and European universities. All the while, the wheels of invention were

↑
The Computer Pioneer Award. The most prestigious prize awarded by the IEEE Computer Society.

1/ The notorious shoe-banging incident occurred during the United Nations plenary meeting on October 12, 1960. One of the delegates was speaking, when Khrushchev took off his shoe, picked it up, and started to brandish it above his head. Contrary to popular belief, he didn't pound the rostrum with his shoe.

At the same meeting, Khrushchev promised to show the United States "Kuzma's mother." What he meant was merely to teach the USA a lesson, but his interpreters, taken by surprise, translated the phrase literally, making it sound incomprehensible and malicious. This incident went down in history, while the phrase "Kuzma's mother" became a Soviet code word for the atomic bomb.

spinning from behind the Iron Curtain in Kyiv's Laboratory #1.

In 1958, the year that Krushchev assumed undisputed leadership of the Soviet Union (and five years after he became First Secretary of the Communist Party), Laboratory #1 was transformed into the Computation Center of the Ukrainian State Socialist Republic Academy of Sciences. The birthplace of the MESM was no longer a humble research lab. More than a rechristening, the change of status meant political importance and with it, more resources. Its new head was Victor Glushkov: a brilliant mathematician, a polymath, and a force to be reckoned with in the field of cybernetics.

Glushkov was born in 1923 in Rostov-on-Don, Russia. But his family soon relocated to Shakhty, a border town between Russia and Ukraine that had witnessed the worst of both Soviet and German power. His youth was bookmarked by bloody shows of totalitarianism. In 1928, the year before the Gluskhkov family moved to Shakhty, the town made international headlines as the site of the notorious "Shakhty Affair." Fifty-three mining engineers and managers were arrested and accused of conspiring with the former mine owners — ousted by the Bolsheviks — to sabotage the Soviet economy. In the show trial that followed, forty-nine men were found guilty and five were eventually executed by firing squad. Glushkov's father, as an engineer, was likely relocated to the town as a replacement for those who had been repressed. Hardly more than a decade later, on June 22, 1941, the Nazi army invaded the Soviet

↓
Victor Glushkov backpacking in the Caucasus, 1955

"I got particularly interested in the 1900 International Congress of Mathematicians where David Hilbert, a famous German mathematician, presented twenty-three of the most complex unsolved math problems. Some of them were solved within the past few years. Someone solving one of Hilbert's problems causes a stir in the scientific community.

"I wanted to work on some understudied area in mathematics, so I picked a very complicated problem pertaining to the theory of topological groups and Hilbert's fifth problem... I wrestled with the fundamental theorem on the fifth generalized problem for three years. My subconscious mind worked even when I was sleeping. I would wake up in the middle of the night, thinking that I'd finally got it. But then I'd get down to work in the morning and I'd find some error or logical inconsistency.

"My three-year battle ended in 1955, when my wife and I went backpacking in the Caucasus. When ice-climbing on a glacier, I suddenly realized how to justify my solution of Hilbert's problem. I was sure, though, that there had to be a logical error somewhere, as usual, so I didn't really believe I'd actually done it. I looked hard for an error, but it was nowhere to be found... I jotted down my solution during the train ride back, and then it took me six months and sixty pages to finalize it. It was a strenuous task, as I was tackling the most abstract areas in mathematics. Most math professors wouldn't even be able to articulate what I'd proven."

Borys Malynovskyi, *Академик В. Глушков: Страницы жизни и творчества* [*Academician Glushkov: Pages of Life and Creativity*], (Kyiv: Наукова думка, 1993).

↓
Victor Glushkov leads a seminar, early 1960s

Union, starting a war to which Glushkov would eventually lose his mother. It was the very day after his high school graduation.

Despite the heavy toll the war took on his family, Glushkov quickly proved himself a promising student. He first enrolled at Novocherkassk Industrial Institute and later transferred to Rostov State University. His first post out of university in 1948 was at a top-secret facility working on a nuclear project as an engineer. Four years later at Moscow State University, he defended his CSc thesis, which proposed a solution to Hilbert's infamous fifth problem.

His solution garnered him acclaim within and without the Soviet Union, launching his reputation as one of the brightest minds of the USSR. Victor Glushkov was just 39 years old when he became the director of the Computation Center of the Academy of Science of the Ukrainian Socialist Republic, but he had acumen surpassing even some of his most seasoned colleagues.

Glushkov had a knack for understanding new ideas, and he threw himself enthusiastically into each of the Computation Center's projects. During the span of his career, he would help pioneer work on artificial intelligence at a time when it was largely the realm of science fiction, winning the Lenin Prize in 1964 and earning himself a seat in the Academy of Sciences of the USSR. He would also go on to develop and present in Munich the world's first machine able to recognize semantic phrases. Under his guidance, the Institute in Kyiv became known as one of the finest hubs of computer innovation not just in the Soviet Union, but in all of Europe.

In his personal and professional life alike, Glushkov was perceived as a giant among men. His wife, at least during their courtship, was nearly adulatory of him. In one letter, she penned, "Dear Victor, after getting to know you, it's been impossible for me to get close to anyone else. You've become my absolute standard. You're an intellectual giant I could never measure up to. I hope you stay like this forever."[2] His colleagues apparently thought more or less the same.

↓
Victor Glushkov and his team, early 1960s. It was Glushkov who put an end to the endless philosophical disputes over the question, "Who's smarter? Humans or machines?" He strongly believed that humans with computers are much smarter and more powerful than humans without them. Another was commissioned by none other than the Joint Institute for Nuclear Research, near Moscow. The Soviet Computer Age had begun.

2/ Borys Malynovskyi, История вычислительной техники в лицах [Pioneers of Soviet Computing], (Kyiv: фирма "КИТ," 1995).

Borys Malynovskyi, an alumni of Lebedev's from the days of Laboratory #1, noted Glushkov's impressive command of not just one speciality, but of all of them. "I was taken aback by his holistic approach," Malynovskyi remembered. "It seemed as if he was observing our world from some position high up above the ground. All the veterans on our team were accomplished scientists, but each of them focused on one specific area, whereas Glushkov could comprehend problems in their totality and give us his insight on how to deal with them."[3]

A true visionary, Glushkov saw a way to turn a fetal computer industry into the lifeblood of Soviet progress. His leadership at the Computation Center would, ultimately, become his life's work. As the Center's chair, he threw himself into cultivating a culture that thrived on intellectual curiosity. Lab assistants and lead scientists alike could find him giving dynamic lectures on the avant-garde topics of the day: logic algebra, automata theory, and new ideas in cybernetics. One-on-one, Glushkov was equally charismatic. Even in chats with colleagues around the center, his passion for a fresh, scientific approach to the world was contagious.

One of his first projects after coming on board was the completion of the Kiev computer, taken over from Sergey Lebedev when he left the lab for Moscow. The Kiev was the first machine in the Soviet Union to use a high-level programming language, Address, invented by physicist and

THE NUCLEAR PROJECT

In 1956, the Joint Institute for Nuclear Research (JINR) was established in Dubna, a "town of science" near Moscow. On February 1, 1957, the JINR was registered by the United Nations. Founded by 11 member states from across the geopolitical world, it focused on nuclear physics and particle physics. Building on the ideas of the "peaceful atom" and openness to collaborating with international colleagues, the Institute aimed to pool the research potential of scientists from all over the world.

The Soviet Union had a 50% share in this project, while 20% was claimed by China. The "nuclear project" participants did study the peaceful uses of nuclear energy, indeed, but military-oriented developments were of higher priority, and the researchers involved in defense industry projects enjoyed more privileges than the others.

The Kiev computer was used, in particular, for remote control of industrial processes, such as heating steel at the Dniprodzerzhynsk Iron and Steel Factory with a telegraph line and an automatic recording unit installed at the site. This helped streamline the process and improve product quality.

3/ Borys Malynovskyi, *История вычислительной техники в лицах* [*Pioneers of Soviet Computing*], (Kyiv: фирма "КИТ," 1995).

mathematician Kateryna Yushchenko. Yushchenko's language remained the go-to language for computing for the next twenty years. One of the machines they designed was installed right there, at the Computation Center, to assist in routine calculations for the lab's work.

The success of the Kiev computer spurred the Computation Center team to green-light the design of a new multipurpose computer intended for industry. This computer would be able to manage production processes in metallurgy, machine building, the energy and chemical sectors, engineering, the food industry, and other fields. The mastermind behind this project was, of course, the indomitable Glushkov, with Malynovskyi as the chief structural engineer.
This incredible computer, christened the Dnepr in Russian, was built in record-breaking time. It took only three years to go all the way from blueprints to fully installed computers at several production sites in 1961.

The Dnepr computer was a small multipurpose machine that featured a special device that enabled it to receive data from industrial enterprises. It was roughly the size of a refrigerator, but it looked tiny relative to the 20-square meter MESM. The machine was also more reliable than MESM because it was based on transistors, not energy-hungry vacuum tubes. And unlike the MESM, the Dnepr didn't need any special cooling or ventilation systems and it could be installed right by the on-site objects of control. It could tolerate a variety of climatic conditions, withstand high humidity, and resist shocks and vibrations.

All of these factors helped make the Dnepr a Soviet computer model with one of the longest production runs. It remained in production for ten years, until it was finally retired in 1971.

The Kiev, MESM, and BESM all had been built part-by-part by teams of scientists specifically for particular projects. The Dnepr was the first computer in the country to be produced on anything larger than a boutique scale. Getting there wasn't easy, though.

To pass government muster, the Dnepr had to be put through three years of testing. First, one machine was presented before a government commission. Then, a couple more had to be installed at several enterprises that used them for almost a year and then decided whether or not to keep them, based on follow-up reports.

After passing that serious round of performance trials, the computers were heavily tested for an additional two weeks. Twenty technicians monitored their performance at room temperature, and then at the thermal load of 30°C above or below freezing (85°F and −20°F, respectively). Next, the computer was given three problems, each related to three different processes: a continuous process — tested experimentally through remote control of the Bessemer converter at the Dniprodzerzhynsk Iron and Steel Factory; a discrete process — tested at the Mykolayiv shipyard; and a learning process — tested at the Kyiv Military School.

Only then was the Dnepr approved for production. Looking back, then-president of the UkrSSR's Academy of Sciences, Borys Paton, called those three years of toil "a heroic epic."[4]

There was not yet a computer manufacturing facility in Ukraine, so the first batch of the Dnepr computers was manufactured at the Kyiv Radio Plant. However, a specialized computer-production facility was not the only thing lacking. So were workers with the skills to assemble computers. The Dnepr may have been the first computer able to be replicated in factories, but that certainly didn't mean it was easily replicated.

The baseline configuration of the Dnepr computer had 2,300 standard cells, 3,000 slots, 23,000 kits, 19,000 solder joints, over 5,000 semiconductor triodes, 120,000 diodes, and more than 150,000 ferrite rings. Overwhelmed with work, the Kyiv Radio Plant manager had hired unskilled workers just out of school for the job. Needless to say, the results were less than spectacular.

The first Dneprs produced were ridden with flaws. Botched solder joints, poorly manufactured circuit boards — every problem a computer could have, these had. Computation Center scientists had to roll up their sleeves and fix the machines. There were only twenty-five technicians to inspect thousands of elements and fix the defects. Overwhelmed, the team had no choice but to bring out the big guns — to send another letter to the Communist Party of Ukraine's Central Committee. The Committee followed up on their request, first enlisting the help of another twenty-five engineers and technicians from the Computation Center's engineering research department.

Glushkov's energy during that period eventually prompted the National Planning Committee to go further and include computing as a goal of its sacred Seven-Year Plan for 1959–1965: a Computing and Controlling Electronic Machines Plant would be constructed before the decade was out (the resulting plant would later be renamed Elektronmash).

4/ Borys Malynovskyi, *Документальная трилогия* [*Documentary Trilogy*], (Київ: Горобець, 2011).

↓
The Computation Center #1 staff working on the Dnepr multipurpose control computer, mid-1960s

This facility manufactured a total of 500 Dnepr computers, which were installed at enterprises and powered industry and science all over the Soviet Union and Eastern bloc countries, among them Hungary, Poland, Romania, and Bulgaria.

The Dnepr was both a symbol and a tool of the Soviet Union's strength. At that time, both the Soviet state and its citizens showed unprecedented enthusiasm for scientific progress. Yuri Gagarin's journey into outer space in 1961 epitomized the Soviet Union's forward march towards progress: it was the triumph of man over nature, of science over religion, of the Worker over the Capitalist. Cybernetics was to play a key role in that mission. Smart people equipped with smart machines would build a brave new world.

The significant calculating power of computers sped up Soviet defense and space exploration technology immensely. Malynovskyi recalls seeing Dnepr's impact at a missile plant in Dnipropetrovsk (present-day Dnipro). "To test its engines," he later recalled in an interview with the authors, "a fifteen-meter tower, as big as an apartment block, was built on the plant's premises. There was a platform with a special rack where the engine was mounted.

"When it started, the entire city knew we were doing the testing. The engine had dozens of sensors attached to it, and a couple of photographers took pictures of them every two or three minutes. The engineers analyzed those shots to get an idea of how the engines worked. This whole process took a few days. But as soon as we configured the Dnepr computer to analyze those sensors, data processing only took thirty minutes."

Elsewhere, the Dnepr computer performed equally well. In Kyiv, the Dnepr was used to remotely control a carbon column producing sodium carbonate some 630 km (390 mi.) away, in Slov'iansk. The economic benefit of the project was so huge that the cost of computer installation was covered in just six months. At an economic cybernetics department, economists used the Dnepr to work out in only three hours the best way to construct a stretch of railway tracks 1,000 km (600 mi.) in length.

The computer-aided scheme not only saved economists and design engineers months of time and effort, but it also shaved twelve percent — 10 million rubles ($9 million at that time, according the official Soviet exchange rate) — off the overall cost. The Dnepr enabled the design team at a synthetic rubber plant to produce a project blueprint in three days instead of the usual forty; at a chemical plant, the design time of new chemical combines was reduced from two months to a week.

Ultimately, though, it was the defense sector and not industry that saw the most benefit from the Dnepr. This computer came with a price tag of 50,000 to 90,000 rubles, depending on the memory capacity, making it far from affordable for most state enterprises. The best customer was the military, which enjoyed nearly unlimited government funding. Malynovskyi recalls that they never even asked about the price.

Enterprise customers, on the other hand, had a harder time. Their budget was regulated by the National Planning Committee, which had sole discretion to determine if this or that enterprise really needed a computer. It goes without saying that its decisions often poorly reflected the actual needs of the enterprise.

Victor Glushkov remained active in computing throughout this time, even as his health began to deteriorate. In 1961, Glushkov published his seminal work, *The Synthesis of Digital Machines*. By this point, he was already deeply engaged in another ambitious project on which, he predicted, the fate of the entire Soviet Union hinged.

His last contribution to science would be equally valued by historians as by scientists. It was a monograph, which Glushkov dictated to his daughter from his hospice bed and published under the unassuming title of *Introduction to Cybernetics*. It chronicled in acute detail his experience navigating the labyrinthine Soviet political landscape in pursuit of scientific advancement.

Glushkov, for all his uncanny mathematical genius, was one of a common genus of Soviet citizen: the indefatigable scientist, determined to usher the Soviet Union into a new era.

↑
The Computation Center, based at the Ukrainian Academy of Sciences' Laboratory #1

"In December 1957, the Academy of Sciences board decided to establish an independent facility — the Kyiv-based Computation Center. They also commissioned an apartment building for its engineers. It was planned that the Computation Center would initially be equipped with three computers: the Kiev model, the SESM, and the Ural-1 model that had just been launched. The facility with three large workspaces could house 400 engineers.

"In 1959, we moved from Feofaniia to Kyiv, even though the center had not been completed yet. That was a very interesting time. According to the technical requirements, the computers had to be installed in clean, air-conditioned rooms, whereas we had no other way but to mount and launch the Kiev computer in a machine room without a roof. It would've been impossible without our incredible team spirit. The facility was completed later, of course."[5]

— From Victor Glushkov's memoirs

5/ Borys Malynovskyi, *Очерки По Истории Компьютерной Науки и Техники в Украине* [*Essays on the History of Computer Science and Technology in Ukraine*], (Київ: Фенікс, 1998).

← An individual core, produced in 1967

→ Different types of vacuum tubes in the 1960s

MAGNETIC CORE MEMORY

Before DRAM (Dynamic Random Access Memory) chips came into widespread use as storage devices in the mid-1970s, computers relied on magnetic core memory, also called ferrite memory. It consisted of an array — or a set of arrays — of ferrite rings strung along a wire. Each ring could represent one bit of data, either "one" or "zero." By today's standards, ferrite memory is bulky, energy-consuming, and labor-intensive.

At the same time, magnetic cores were resistant to nuclear and electromagnetic radiation, unlike modern microchips. Given that, ferrite memory was installed in military and other special-purpose computers for decades. Up until 1991, it was even used in onboard computers of American space shuttles.

VACUUM TUBE VS. TRANSISTOR

In the 1940s–1950s, the vacuum tube was the main building block of an electronic computer. It was invented back in 1883, by the famous American inventor Thomas Alva Edison, and rather unexpectedly at that. Edison was trying to increase the service life of a regular lamp, and during one of his experiments, he inserted a metal plate with a conductor positioned outside into a vacuum flask.

It turned out that vacuums conduct electric current, but only one way — from the cathode to the anode — and only if the cathode is hot. This phenomenon is known as thermal electron emission.

It was a startling discovery because at that time, it was believed that vacuums were not able to conduct electric current, as they didn't have any charge carriers.

052

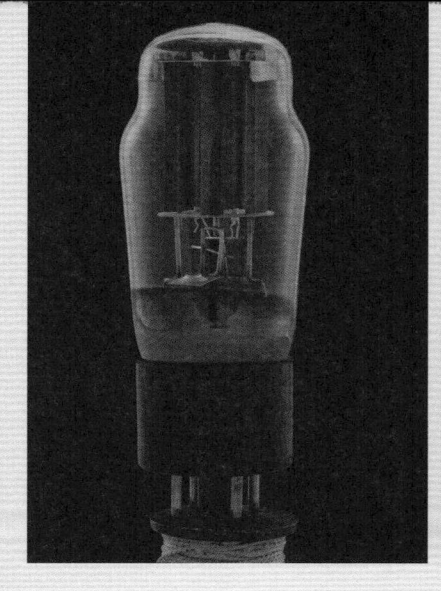

Although oblivious to the implications of his world-changing invention, Edison went ahead and patented it. In 1904, the Edison Effect formed the basis for another patent granted to Sir John Ambrose Fleming for a device converting the alternating current produced by radio waves into direct current that he called the oscillation valve. This diode rectifier (also called a hot-cathode tube, tube diode, kenotron, thermionic diode or the Fleming valve) ushered in a new era in electronics. Two years later, American engineer Lee de Forest added the third electrode, the grid, between the other two, inventing the triode that could operate as an amplifier.

Vacuum tubes needed extra energy to heat up the cathodes and derive cathode emittance. So, first-generation computers had drawbacks — they consumed too much electricity and had an overheating problem. Moreover, vacuum tubes, just like regular lamps, often blew out, causing hardware interruptions.

There is often a flashed coating on the inner surface of vacuum tubes. It's called a "getter," an absorber of small amounts of gas left inside during assembly. Some types of getter would turn white, when exposed to to air, indicating a failure of a hermetic seal.

Second-generation computers, more compact in size and reliable, emerged in the early 1950s, with the development of a transistor, a semiconductor triode that replaced vacuum tubes. The transistor (whose inventors William Bradford Shockley, John Bardeen, and Walter Houser Brattain were awarded a Nobel Prize) was small, lightweight, and cheap. It was also resistant to physical shock, energy-efficient, and durable.

Transistors didn't heat up as much as vacuum tubes, which solved the problem of heat dissipation. Neither did they need to warm up the cathode, so the computers started faster. New computer generations followed, but none of them created such a seismic shift as the transition from the first, tube-based generation to the second, transistor-based one.

← Transistor, late 20th century

THE ANATOMY OF A TRANSISTOR

A transistor is a simple device composed of multiple layers of silicon with different physical properties and with at least three aluminum terminals. Simple as they are, transistors are the foundation for the most advanced electronic devices.

In simple terms, they allow electronic devices to be smaller, enabling inventions like walkie-talkies (used by American soldiers in World War II) and our constantly shrinking televisions and mobile phones today.

A transistor acts as a switch — it allows electrons to flow from one terminal (source) to another one (drain) and then interrupts the flow when necessary. The difference is that transistor is switched on or off not mechanically, like a regular switch, but with a small current applied to the third terminal (gate).

To switch a transistor on, a small positive charge is applied to the gate and then transferred to the conducting polysilicon layer isolated from other parts of the transistor with silicon carbide. The positive charge, stored in the polysilicon layer, attracts negatively charged electrons from the p-type silicon base that separates two n-type layers adjacent to the source and the drain.

Electron outflow results in an electric discharge in the p-type silicon compensated by the inflow of negatively charged electrons from the n-type silicon layer adjacent to the source. These electrons not only fill in the vacuum but also flow to the n-type silicon region adjacent to the drain. So, the negatively charged electrons start moving from the source to the drain — a transistor is on and stores a "one" (bit 1). If a small negative charge is applied to the gate, the negatively charged electrons repel each other, so the electrons no longer flow from the source to the drain, and the transistor switches off, reading back as a "zero" (bit 0).

When the transistor finally appeared — 1947 in the U.S. and 1949 in the USSR — the achievement was accurately understood as the start of a technological revolution. In both the Soviet Union and the United States, the invention of the transistor was preceded by decades of research and experiments. Six years before the Bell Telephone Laboratories invented the transistor in 1947, Ukrainian scientist Vadym Lashkaryov accidentally discovered the p-n junction. This was a pivotal step, leading the way to the first semiconductor diodes, which were used by the Soviet army in World War II.

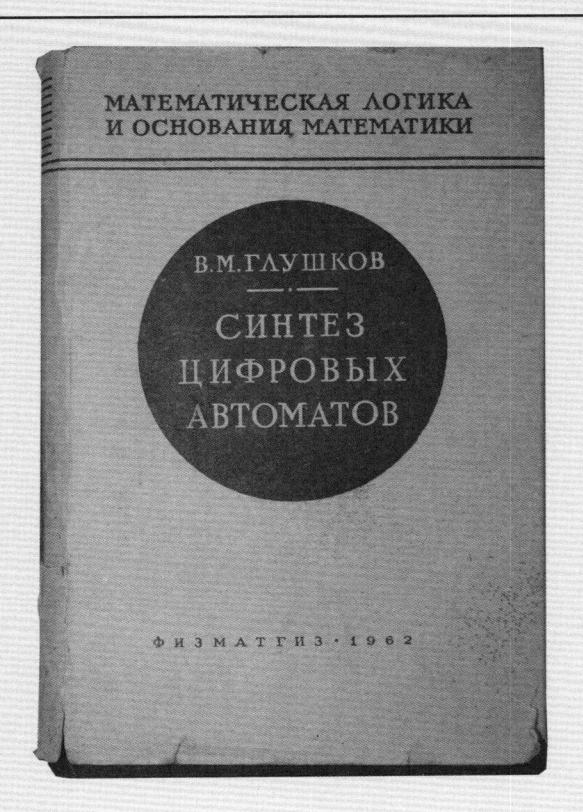

Then in 1949, around the same time that Glushkov's team at Laboratory #1 was preparing to make history with their MESM computer, Kyiv Polytechnic Institute graduates Oleksandr Krasylov and Susanna Madoyan succeeded in building the first working prototype of a transistor in the USSR. Krasylov moved on to chair a research lab developing the first semiconductor devices in the Soviet Union, including a two-dimensional germanium transistor (a predecessor to today's ubiquitous silicon transistors) in 1953.

The Synthesis of Digital Machines by Victor Glushkov was the first book of its kind, with an original run of 15,000 copies. Targeted at researchers, structural designers, and engineers, the book covered "the issues of rational design of program-controlled electronic digital machines."

←
The Synthesis of Digital Machines by Victor Glushkov, published by the Physics and Mathematics Literature Press in 1962

Scientists of the Academy of Sciences of Ukrainian SSR checking the work of Dnepr-2

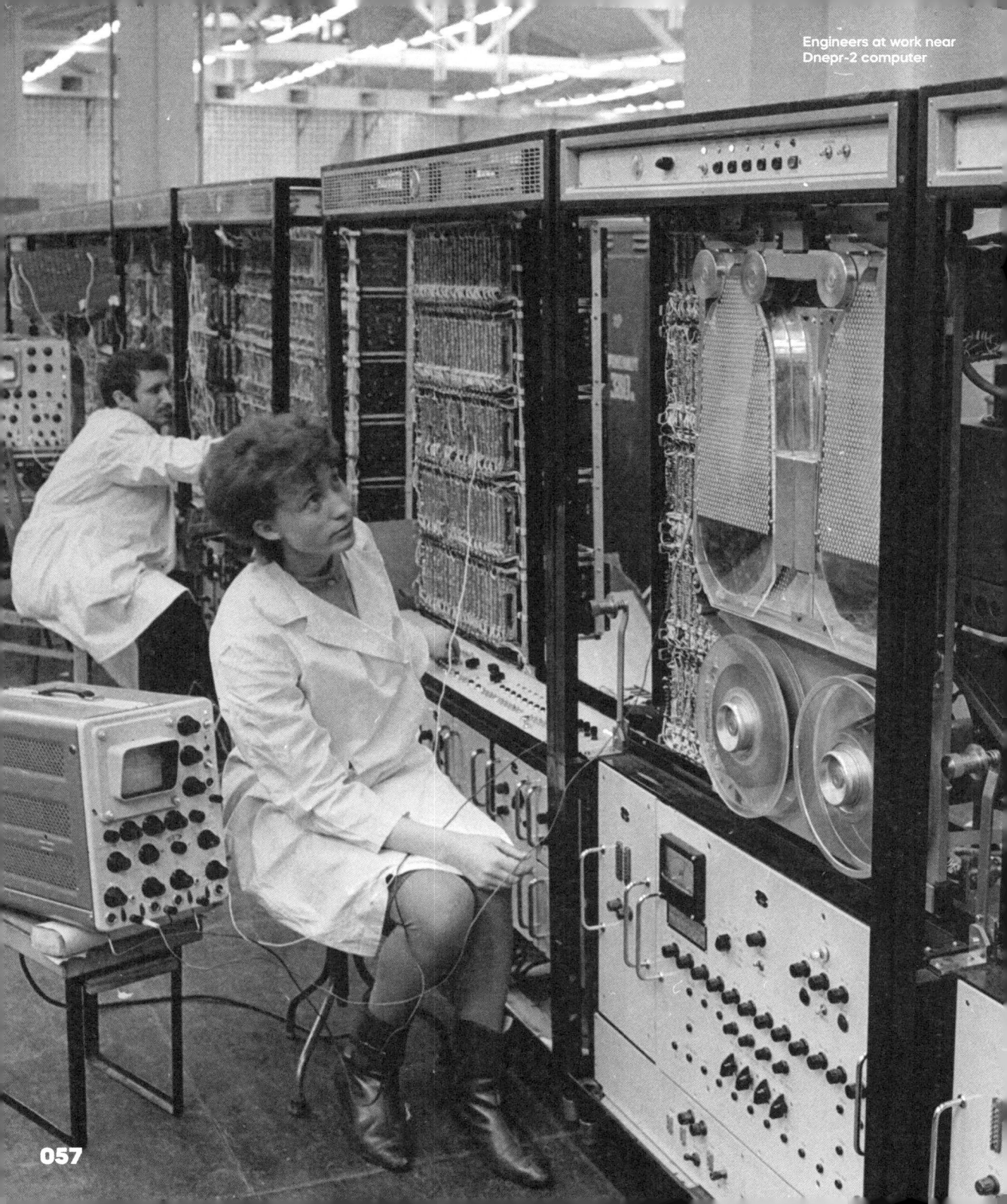

Engineers at work near Dnepr-2 computer

chapter 4

↑
Area for setting up computers at the
Kyiv Plant of Computing and Controlling
Machines (now Kyiv's Electronmash
research and manufacturing association)
in 1969

The Heyday of Cybernetics in the USSR

The USSR in the 1960s was alight with the bustle of industry and innovation. New factories transformed bleak towns into busy cities, millions of Soviet citizens moved into their own apartments, and a young pilot from Smolensk Oblast, Russia, became the first human to go to space. In the Kremlin, Nikita Krushchev had replaced Stalin in leadership, ending the Reign of Terror that had stymied the field of cybernetics in the days of Laboratory #1. Optimism was in the air.

Nonetheless, bureaucratic traditions died hard in the Soviet Union. Even if Stalin himself was dead, the system he set in place was not. In the early days of the country, Stalin maintained his authority by establishing a rigid hierarchy in which one person, the general secretary of the Communist Party, exerted nearly absolute power. This power structure was replicated in all professional fields.

For instance, in the field of rocket and space research, key decisions were made only by Chief Designer Serhiy Korolev; in the field of nuclear research and the creation of the atomic bomb, the person with the deciding vote was the bomb's creator, Igor Kurchatov; and in publishing, the Chairman of the State Committee for Publishing, Printing, and Book Trade served as gatekeeper to every publishing outlet in the Soviet Union. And so things were in absolutely all areas of life. While leading roles in any industry could count on a higher quality of life than a worker or low-level bureaucrat, it was the top experts who claimed the full blessings of the Soviet elite. No makeshift labs and dormitories for them — these leaders had access to nearly unlimited funding and resources for their projects. And the benefits didn't stop at the office door. The government took good care of them, offering perks like access to restricted-access grocery stores with products not available to most Soviet citizens. They could also expect generous rewards for good performance. Kurchatov, for example — the creator of the Soviet atomic bomb —received a handsome bonus in addition to his government salary. On top of this, he was given his own mansion and country house — complete with a maid staff — a luxury car, free public transportation passes for him and his family, and a doubled salary during the rest of his tenure as the head of the Soviet atomic program.

Competition for these coveted positions was fierce, and people played dirty to win. Some would stoop to slandering their rivals or writing letters of denunciation just to get ahead. In this environment, it was difficult to win out on merit alone over those willing to utilize political means and demonstrations of loyalty to the Party. Computer science was no exception. Eager to take advantage of the government's new position on cybernetics, Lebedev's colleagues at the Institute of Mathematics in Kyiv went head-to-head with Glushkov to lay claim to the field (and its potential rewards).

In the long run, the accusations against Glushkov petered out for lack of an audience. Within three years, separate departments for theoretical, technical, biological, medical, and economic cybernetics had opened their doors.

Throughout the 1960s, the Institute of Cybernetics focused a great deal of its energy on developing industrial control systems for state industries, led by Glushkov. He was the ultimate multitasker. He was all-in on the immediate projects at hand, without losing sight of his long-term dream: OGAS, an all-union industrial control system and prototype for the internet in the Soviet Union. OGAS would take him from Kyiv to Lviv and then the Kuntsevo Radio Factory outside of Moscow, where he created small-scale prototypes of industrial control systems in the 1960s and early 1970s.

He eventually built a more comprehensive industrial control system that encompassed computer-assisted design, technology management, product testing, and managerial control for an aircraft factory in Ulyanovsk, Russia. But the full saga of OGAS wouldn't be told until 1982, when Glushkov lay on his deathbed.

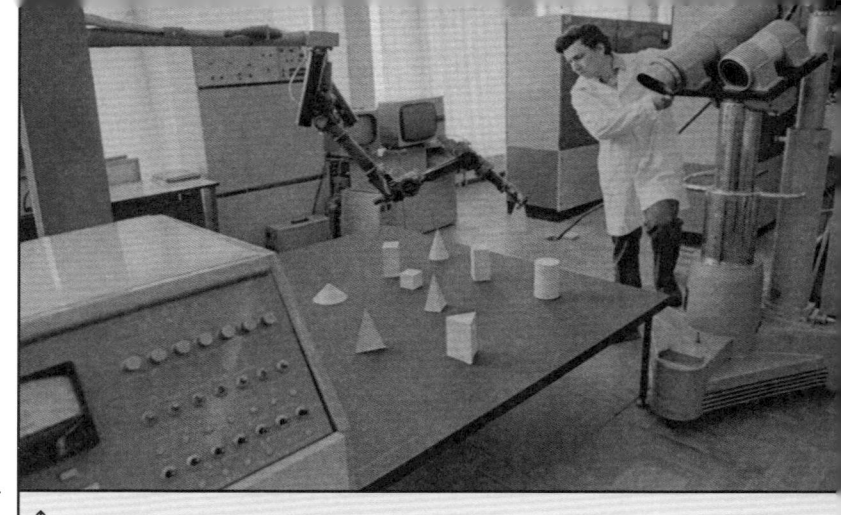

↑
Viktor Lanbin, senior engineer at the Institute of Cybernetics configures the Eye-Hand system of the central robot, Kyiv, 1978. In the 1970s, the Cybernetics Institute pioneered mind process modeling, including the Eye-Hand system. This innovative setup featured a TV camera and a 6-degree-of-freedom manipulator controlled by the M6000 computer. The powerful BESM-6 processed images, recognizing geometric objects and planning their manipulation, showcasing advancements in self-learning image recognition.

At that time, the Institute of Cybernetics began developing a new computer for engineering calculations. In 1963, Promin, the forerunner to the personal computer, saw the light of day. According to Glushkov, it was "a groundbreaking invention that had several hardware advances, including metallized memory cards." Moreover, it was the first widely used computer with step-by-step microprogramming control. This system wasn't patented, but Glushkov did receive an inventor's certificate in the USSR. Subsequently, this same system was used for MIR-1, an early Soviet personal computer.

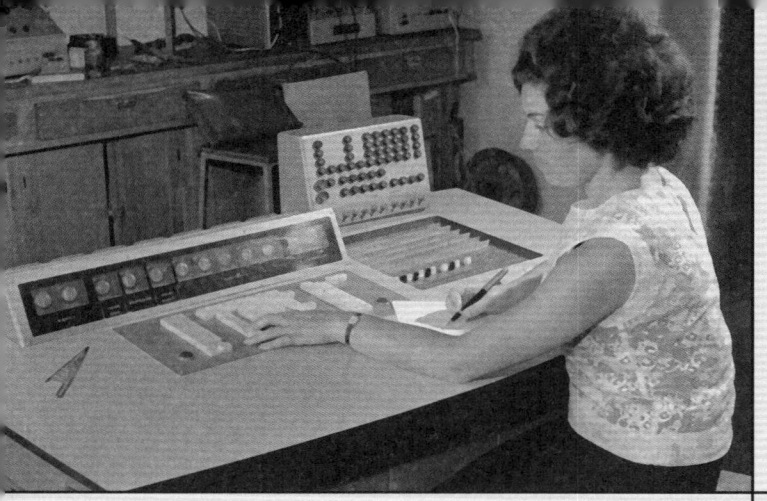

← Electronic calculating machine "Promin" ("ray of light" in Ukrainian) at the Institute of Cybernetics of the Academy of Sciences of the Ukrainian SSR, Kyiv, 1982.

↓ Victor Glushkov in the late 1960s with Yuliia Kapitonova, an Institute of Cybernetics alumna. Kapitonova would go on to win numerous accolades for her pioneering work on digital automation, including the State Prize of the USSR (1977) jointly with Glushkov.

In 1963, Kyiv hosted a computing conference for participants from Bulgaria, Czechoslovakia, Hungary, and Poland. Later that year, the Kyiv-based Institute of Cybernetics and Uzhhorod State University organized a small symposium in the Transcarpathian city of Uzhhorod. Sergey Lebedev himself attended the symposium and hammered out a deal with Glushkov. From then on out, Lebedev's team at the Institute of Precision Mechanics and Computer Engineering in Moscow would develop large computers (the BESM series, in particular), while Glushkov's institute would handle more specialized machines.

The Uzhhorod Symposium inspired him so much that afterwards, he developed a comprehensive plan for the MIR computer in just two weeks' time. Many of Victor Glushkov's colleagues and protégés admired his tremendous work ethic, enthusiasm, and bursts of inspition.

← A programmer at work on the MIR computer

Glushkov once wrote: "When designing the MIR computers, we set ourselves an audacious goal — make machine language as close to human language as possible. I'm referring to mathematical, not spoken language, although we've experimented with creating computers that can produce human language."

"In cybernetics, sometimes coming up with an idea is just 0.01% of the equation, and making it a reality is the other 99.99%. That's why synchronizing your short-and long-term goals is such a crucial management principle. Basically, this means that in a new field, like cybernetics, you shouldn't focus on achieving a concrete, short-term goal if you can't see how the project will progress beyond it. On the other hand, you should never undertake a long-term endeavor without considering whether you can break it down into individual phases or not. Each phase should be a step towards your major goal; meanwhile, it should also provide some sort of concrete benefit and feel like an achievement in its own right."

— Victor Glushkov[1]

1/Malynovskyi, *История вычислительной техники в лицах* [*Pioneers of Soviet Computing*].

↑
Glushkov giving a presentation at a conference
in Lviv, 1965

CYBERTONIA

After Stalin's death, Nikita Khrushchev questioned his leadership and denounced his "cult of personality." Soviet citizens were given a new set of moral standards. In the 1960s, the public held scientists in high esteem, and they were chosen as role models. Simply put, they personified integrity because absolute truths — two plus two is always four; gasses expand when heated — formed the basis of their professions. Scientists embraced their new role. As Pyotr Kapitsa, a leading Soviet physicist and a Nobel laureate, once wrote, "In order for the democratic process and the rule of law to work, every country must have independent institutions that serve as arbiters in constitutional issues.
In the USA, the Supreme Court fulfills that role, while the House of Lords does in the United Kingdom. In the Soviet Union, it appears as though that moral function has fallen upon the Academy of Sciences."[2]

2/ Alexander Genis and Peter Weil, *60-е. Мир советского человека* [*The 60s: The World of the Soviet Man*], (Moscow: ACT, 1998).

An atmosphere of hope, along with an approximation of creative freedom, characterized the USSR in this period. In keeping with the spirit of the times, young researchers of the "Thaw generation" were eager trend-setters inside the lab and out. They played in bands, performed in comedy shows, organized fun camping trips, and even "discovered" their own country — Cybertonia.

↑
Passport of the citizen of Cybertonia

Cybertonia first appeared on the map of Kyiv on December 31, 1962. Leonid Sapozhnikov, who worked at the Institute of Cybernetics in his younger years and later became a well-known writer and journalist, conceived this wonderful country. Incidentally, almost all the employees at the Institute were young — the average age was just 23 years old. They were ecstatic about the sign hanging by the director's office on the second floor: "Dreamers and artists have 'discovered' a new country — Cybertonia. You can't see it with your eyes, and it won't fit in the lens of any binoculars, because it's a country that exists in four dimensions: energy, laughter, dreams, and fantasies!"[3]

3/ Ihor Osipchuk, «Академик Глушков от души смеялся, посмотрев пародии о своем институте кибернетики» ["Academician Glushkov Laughed Heartily After Watching Parodies About His Institute of Cybernetics"], OGAS. January 9, 2014, http://ogas.kiev.ua/ua/perspective/akademyk-glushkov-ot-dushy-smeyalsya-posmotrev-parodyy-o-svoem-ynstytute-kybernetyky-766.

Cybertonia was the place to be for New Year's in 1964. The Institute had to rent out the October Palace, a performing arts center that could seat over 2,000 spectators, to accommodate all the party guests. Naturally, not everyone could be invited, so there were people trying all sorts of schemes to get in — even making counterfeit tickets! And these tickets were quite special. They looked like passports for citizens of Cybertonia, complete with the country's constitution written on it. "It is every citizen's honorable duty to be in a good mood" read one of the clauses. Cybertonia even had its own currency, the cybertino, which you could spend on sparklers, tinsel, party poppers, masks, and other knick-knacks at Cybermart.

Invitees couldn't just waltz through the front door of the palace, though. They had to worm their way through a hole in the fence to enter the fantasyland. Cybertonia's popularity extended well beyond these New Year's parties — its merry citizens performed at various other institutes, traveled to other cities all across the Soviet Union, provided entertainment at children's birthday parties, and printed their own newspaper. This hopping fantasyland was brimming with real-life adventures and fascinating events.

Cybertonia even had its own civil registry office that issued couples marriage certificates on punch cards together with two copper rings. Numerous competitions were held; there was one witty performance after another, including satirical skits in which junior employees poked fun at their bosses. Robots took care of everything in Cybertonia, obviously! Many young people showed up to the party dressed as robots.

At the end of the evening, guests elected two monarchs to rule Cybertonia — Don Cyberton and Signorina Cybertina. The candidates took part in goofy little competitions, and the participants for whom the audience clapped the longest were declared the the winners. The judges even timed everything with stopwatches. Could anything different be expected from science geeks?

GENERATIONS OF COMPUTERS

The history of computers can be divided into several generations. First-generation, vacuum tube computers, such as the MESM, the SESM, and the Kiev computer, which reigned in the 1950s, were enormous and remarkably slow, yet not particularly reliable. They had a small operating memory and very few accessories. There were no interrupt systems or multiprogramming capabilities. They contained only a few components of modern translation and operating systems.

In the 1950s and 1960s, second-generation computers quickly rendered vacuum-tube computers obsolete. These were transistor-driven machines that were more compact than their predecessors. Transistors made central processors faster and significantly more reliable, and cut down on energy consumption. Also, the core memory could store more data, and the machine was faster. These computers used faster peripherals and more sophisticated programming languages. Scientists developed various rudimentary components of parallel processing, conducted experiments involving multiprogramming and time sharing, and created rather satisfactory operating systems, too.

High-level programming languages, such as Fortran, Algol, Cobol, and others, appeared on second-generation computers like Dnepr and MIR-1, which meant that programs no longer had to be tailored to a particular computer model. From 1965 through 1967, several companies outside of the Soviet Union, IBM being the largest of them, manufactured third-generation computers, which subsequently replaced second-generation models.

Third-generation computers were built using integrated circuits. Before they were invented, each logical unit designed for remembering and transforming each bit of information consisted of several physical units (transistor diodes and triodes, resistors, capacitors, etc.). Integrated circuits, on the other hand, were one physical unit that encompassed several logical units.

This generation of computers were based on the building block concept and certain devices functioning autonomously, which paved the way for more developed operating systems that allowed for more efficient use of all the computers' capabilities. Computer architecture included communications processors and standardized interfaces with peripherals.

Fourth-generation computers, which first appeared in the early 1970s, were equipped with large integral circuits. We still use fourth-generation computers today. In 1972, Victor Glushkov clearly saw where the industry was heading, and he even predicted what features computers of the future would have. Incidentally, the word "computer" became part of the average person's vocabulary at around that time. These computers were either classified as supercomputers, like Elbrus, manufactured in the Soviet Union, or as PCs, such as Iskra, Elektronika, and others.

Researcher at the Institute of Cybernetics of the Academy of Sciences of the Ukrainian SSR at the new electronic machine "Promin", Kyiv, 1982.

PROMIN

In the Soviet Union, the Promin ("ray of light" in Ukrainian) computer was designed to perform engineering calculations well before programmable calculators were made available. The Promin and other similar machines were created under Victor Glushkov's leadership in 1963. Capacity-wise, this model couldn't compete with the BESM, since it was supposed to be used for making on-the-spot calculations. Promin wasn't a bulky machine. Roughly the size of a desk, it could perform approximately 1,000 addition operations (8-10 times fewer than the BESM) or 100 multiplication operations per minute. A keyboard sufficed for typing out the numbers, but special jack plugs or metallized punch cards were needed to input the commands.

Promin could solve differential and linear equations, calculate the extrema of nonlinear functions, etc. Promin-M and Promin-2, updated, more powerful, and more functional models, appeared in 1965 and 1967, respectively. Promin, as well as the MK-52, a programmable micro calculator designed to solve engineering calculations, are on display at the State Polytechnic Museum in Kyiv. In the late 1980s, the MK-52 flew to space on the *Soyuz* TM-7. If the onboard computer had failed, this micro calculator would have calculated the landing trajectory.

Although both machines performed similar functions, the MK-52 was significantly smaller. This third-generation calculator took four AA batteries, weighed 250 g (8.8 oz.), could use a memory expansion unit, and it cost the equivalent of $191 at the time ($641 in today's money).

068

МК 52
СМ.ПерМ
СМ.ПерП
Зн.РгП
РгП
Зн.РгМ
РЕГИСТР МАНТИССЫ
ТОп
РЕГИСТР ОПЕРАЦИИ
Зан.в См.
СЧЕТЧИК МИКРОКОМАНД
Вкл.
16
8
4
2
1
8
4
2
1
ОСТАНОВКА ПО СЧЕТЧИКУ МИКРОКОМАНД
N КОМАНДЫ
НАЧ.СБРОС
ПУСК
ОСТАНОВКА
Вв1
Вв2
Авт
Такт
К1
К2

MIR

In 1965, MIR (short for Machine for Engineering Calculations) replaced Promin. (Coincidentally, the abbreviation is also the Russian word for "world.") A Soemtron electronic typewriter, manufactured in East Germany, was used to input data. The output was shown on a tube display. Thanks to this convenience, it was quickly slated for mass production for research institutes and engineering companies.

MIR had one remarkable feature — its hardwired machine language that was close to high-level programming languages. In 1968, a more modern model equipped with an eight-track tape came out, the MIR-1. MIR performed calculations much faster than Promin.

↑
Engineers from the Laboratory of Computer Devices at the Cybernetics Institute are analyzing the progress in solving a problem on the Mir computer in Kyiv in 1965.

It could solve 200–300 operations with 5-bit numbers per second. Its core could store 4,096 12-bit words, it weighed 400 km (881 lbs.), its power supply was 380 W, and it consumed 1.5 kilowatts per hour.

MIR-2

By the late 1960s and early 1970s, computers were starting to work more and more like they do today. Although inventions like the mouse and the graphic pen didn't catch on until the 1980s, graphic displays were used quite often. They were vector displays, not pixel-based ones, which meant that every symbol and every one of its lines were traced on the inside surface with a beam.

Victor Glushkov was, predictably, one of the first to write about these input and output tools. In his 1972 book *An Introduction to Automated Control Systems*, he describes the early forerunner of modern-day computer interfaces: "There are special devices for inputting graphics — sketches and outline pictures — into a machine by circling all the lines with a special pencil on a special tablet […].

"Displays provide vast opportunities for fostering a dialogue between humans and computers. Displays look like televisions screens, and computers can put letters, numbers, or graphics on them. By using a special device called a "graphic pen," people can draw on the screen or erase images, move symbols or parts of a picture, put new signs on the screen, on spots marked by the graphic pen, etc."[4]

MIR-2, first manufactured in 1969, was equipped with both a vector display and a graphic pen. This made it similar to today's personal computers. MIR-2's main features included roughly 12,000 operations per second, 8,000 13-bit word operational memory, external memory on perforated tape, and magnetic cards.

In 1971, a new, tremendously more powerful computer came out, the MIR-3. However, it never enjoyed the popularity its predecessors did.

4/ Victor Glushkov, Введение в АСУ [Introduction to Automated Control Systems] (Kyiv: Техника, 1974).

IMAGE RECOGNITION AND THE COORDINATION SYSTEM *EYE-HAND*

In the 1970s, the Institute of Cybernetics actively tried to model thought processes. In particular, it focused on image recognition. "Universal digital machines make it possible to model and test various ways of recognizing images, including machine learning systems," Victor Glushkov wrote in his work *Cybernetics: Theory and Practice*. "As of today, numerous such systems have been developed and tested. For instance, the system learning to recognize shapes imitates the adaptive traits of the human brain in terms of that particular Activity."

← Vasiliy Krot, team leader at the Institute of Cybernetics.
Kyiv, September 1982.

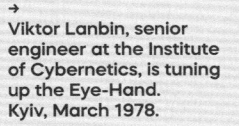

Georgy Gimel'farb, a contemporary of Glushkov at the Institute, shared his thoughts on the photos (right, below). "These photographs were probably taken for a newspaper or magazine article. They show Eye-Hand, a coordination system built with the support of Victor Glushkov. The system consists of a TV camera on a movable platform and a six-degree-of-freedom manipulator controlled by M6000, a small computer."

→ Viktor Lanbin, senior engineer at the Institute of Cybernetics, is tuning up the Eye-Hand.
Kyiv, March 1978.

The camera and M6000 were hooked up to BESM-6, a powerful machine that captured the image of a desk with basic geometric objects — cubes, wedges, and pyramids — on it and processed it, detecting and recognizing these objects. Then it planned the acquisition trajectory and movement of individual objects to different places.

"I have no idea why the photographer had Vasiliy assume such an awkward pose. He probably wanted to emphasize just how similar humans and robots are in certain ways."

"Lanbin's picture shows nearly the whole system: the control unit of M6000 in the bottom left corner, the manipulator in front of the desk with the objects on it, Lanbin manning the camera, and part of BESM-6 behind him."

↑
MIR-1 computer exhibited at Borys Paton State Polytechnic Museum in Kyiv, Ukraine

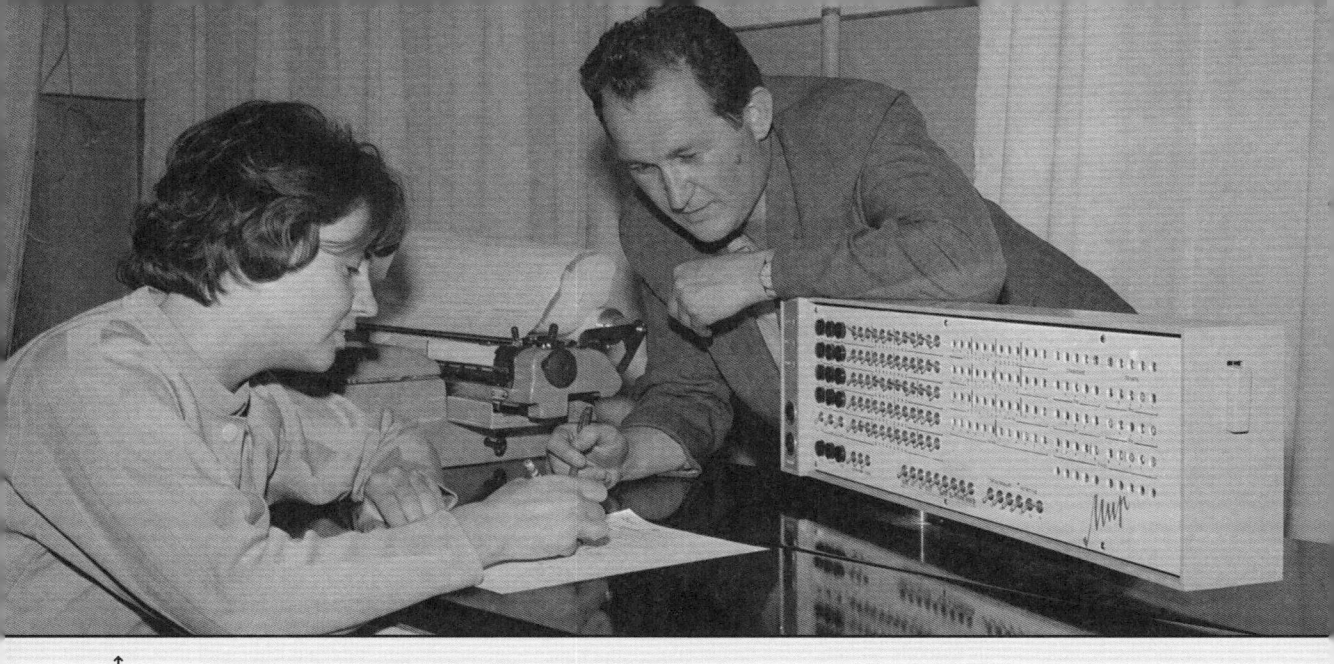

↑ Engineers at the control panel of the new MIR computer

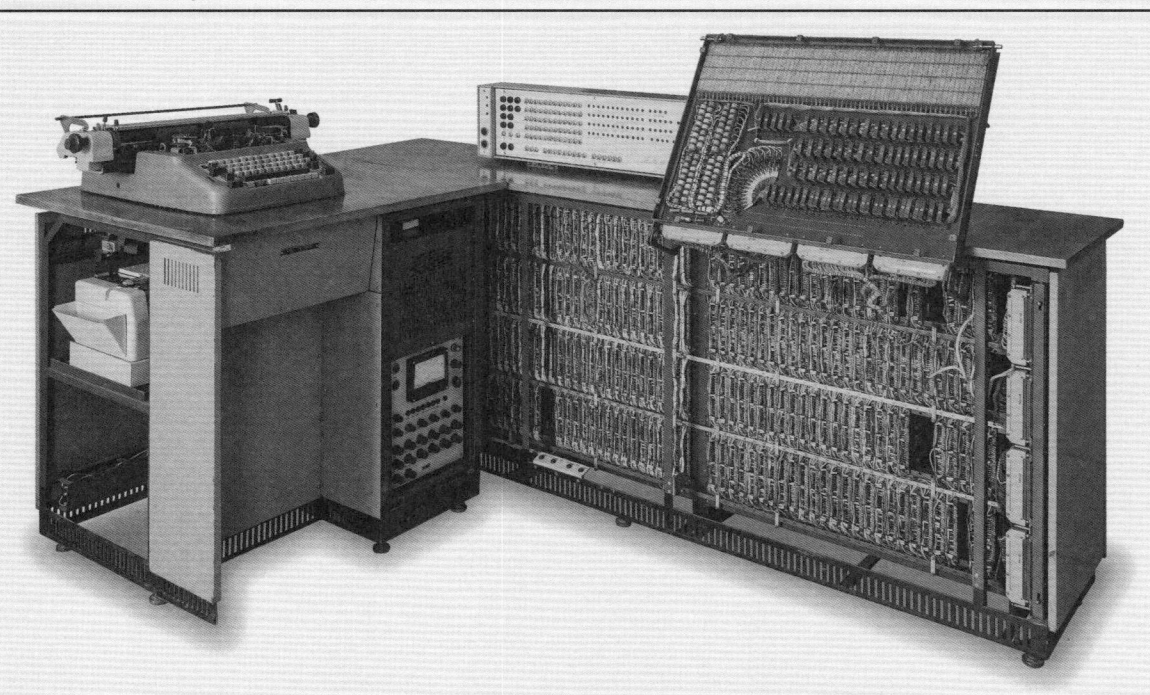

← A data entry operator types information into the MIR-1 computer

↑ Inside the same MIR-1 computer exhibited at Borys Paton State Polytechnic Museum

074

Control computer complex M4030 developed in 1973 at the Kyiv Research and Production Association "Electronmash"

076

МИР

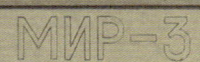

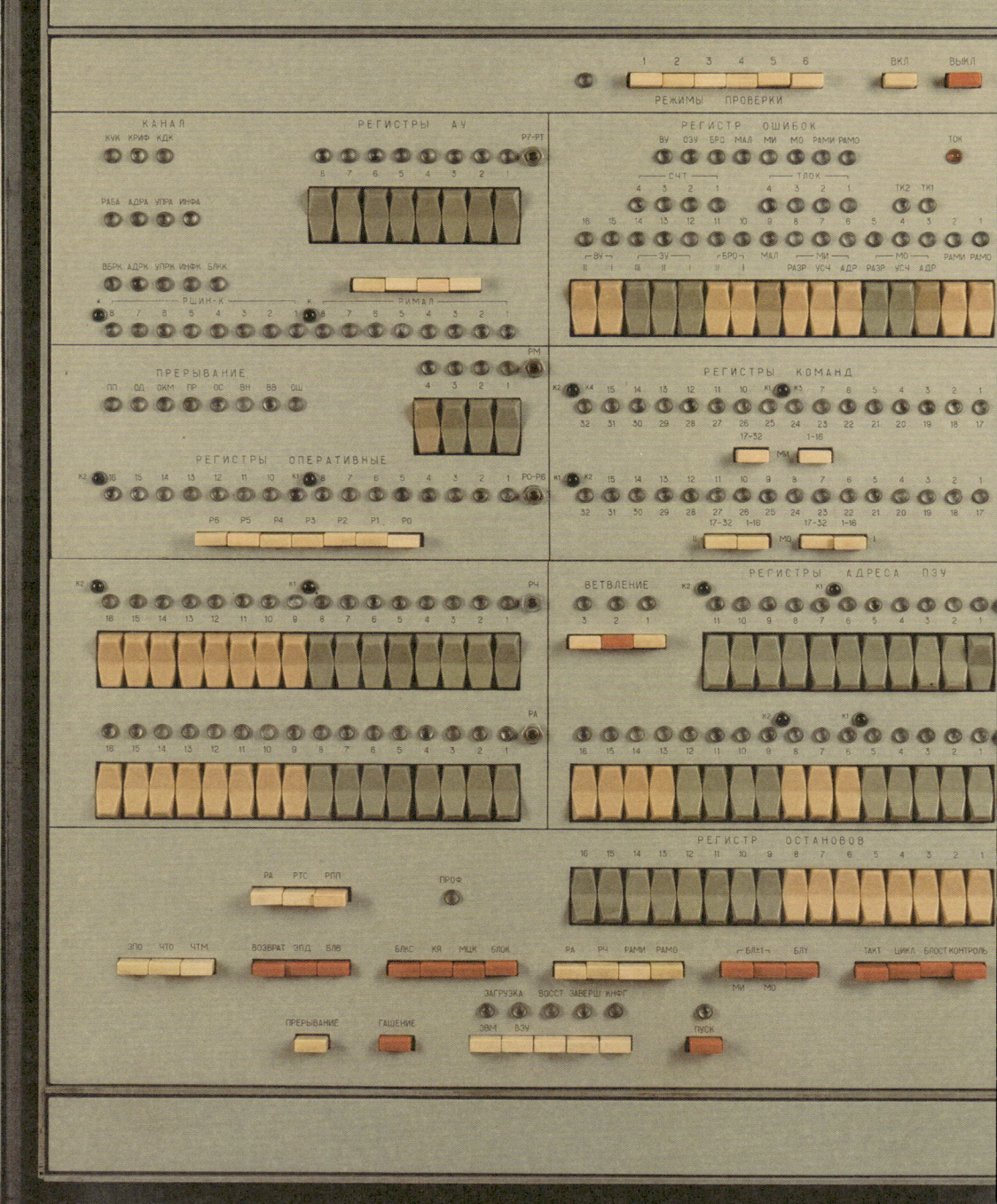

MIR-3 computer exhibited at Borys Paton State Polytechnic Museum in Kyiv, Ukraine

| 16 | 15 | 14 | 13 | 12 | 11 | 10 | 9 | 8 | 7 | 6 |

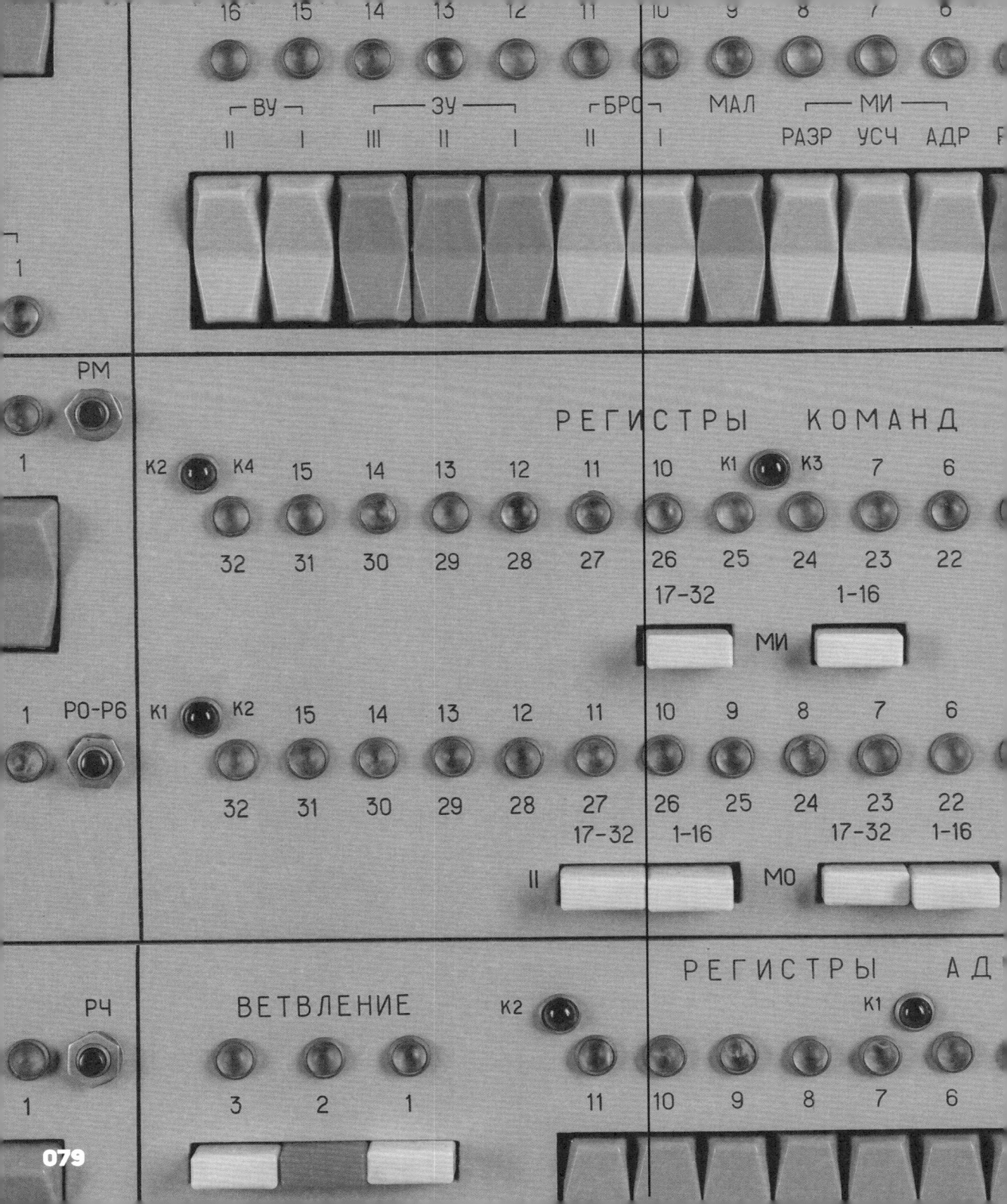

┌─ ВУ ─┐ ┌─── ЗУ ───┐ ┌─ БРО ─┐ МАЛ ┌── МИ ──┐
   II   I   III   II   I   II   I   РАЗР УСЧ АДР

PM
1

РЕГИСТРЫ КОМАНД

К2	К4	15	14	13	12	11	10	К1	К3	7	6
	32	31	30	29	28	27	26	25	24	23	22
							17-32			1-16	

МИ

P0-P6
1

К1	К2	15	14	13	12	11	10	9	8	7	6
	32	31	30	29	28	27	26	25	24	23	22
						17-32	1-16			17-32	1-16

II   M0

РЕГИСТРЫ АД

РЧ   ВЕТВЛЕНИЕ   К2   К1
1   3   2   1   11   10   9   8   7   6

079

BESM-6 was the first Soviet second-generation supercomputer, widely used for general computation and control tasks

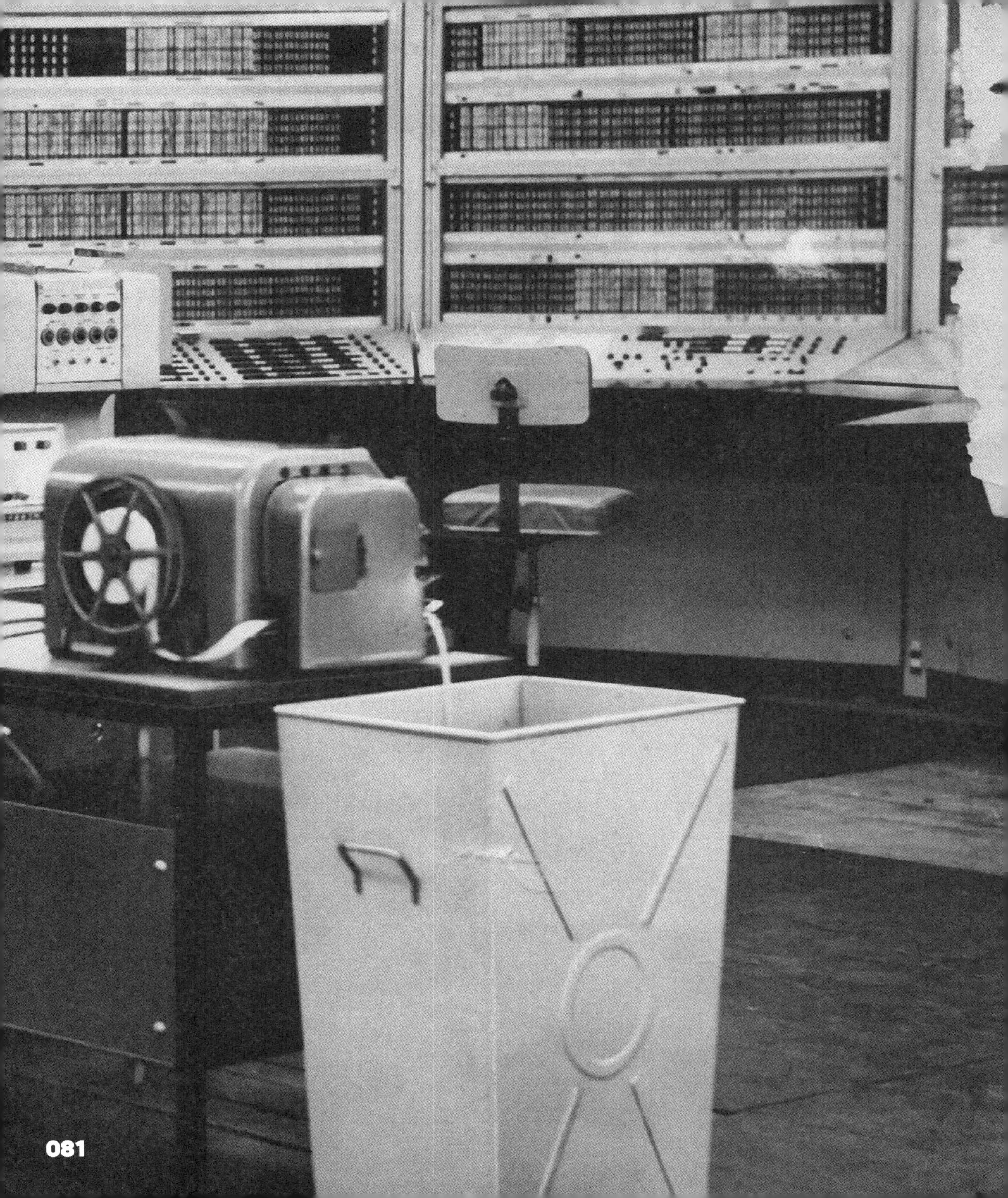

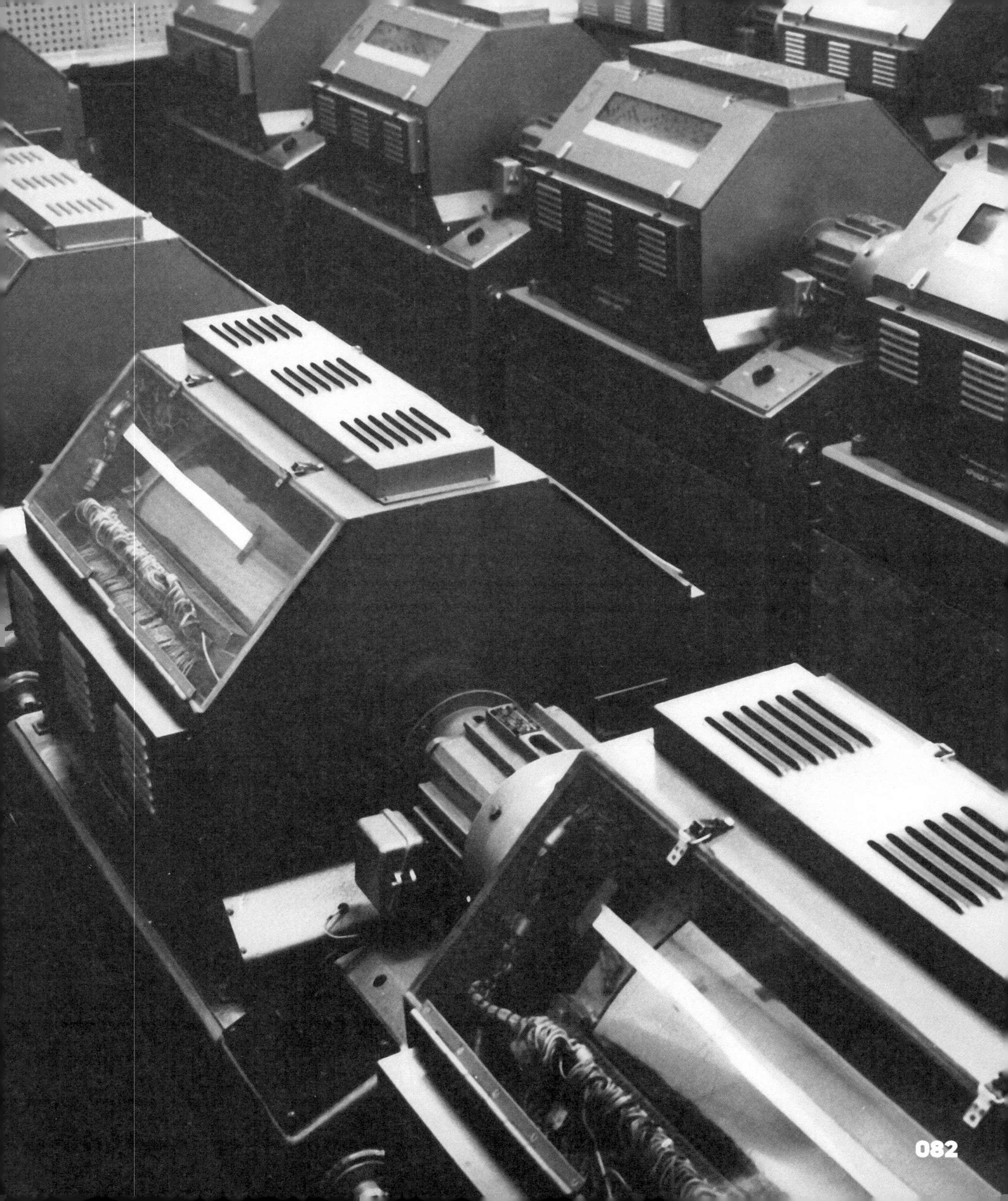

↑
Engineer at work on BESM-6

←
Up to 16 magnetic drums with a capacity of 32K words each (200 KB) were used as external storage for BESM-6

chapter 5

↑
The Computing Center of the Soviet Academy
of Sciences' Institute of Cybernetics

OGAS: The Soviet Internet

In the early 1960s, Victor Glushkov set out again to revolutionize the Soviet Union. It wasn't a revolution of ideals he was after. Glushkov's was a revolution of efficiency. His enemy #1 was the arbitrary, error-laden bureaucratic procedures that were slowly but surely setting the Soviet economy on a destruction course. His weapon would be a complex network of computers, connecting production sites to each other nationwide via a broadband cable — the Soviet Internet.

Victor Glushkov was a futurist before the term was in vogue. He came of age in a Soviet Union still reeling from World War II. By the time he had begun his career in earnest, however, both population and production were again growing rapidly. To many in the Soviet administration, this looked like a good thing, but Glushkov knew better.

As a mathematician, he saw the world in equations and solutions. Knowing first hand the mess of clunky, error-ridden administrative procedures governing the Soviet Union, he predicted that the economy could not be scaled much further. If something didn't change, then "by 1980," he would later predict, "the entire Soviet adult population would have to be engaged in planning and management."[1]

When Glushkov first heard in the 1950s about the emerging field of cybernetics, he decided it was the solution he was looking for. In 1956, Glushkov began devising a plan for the National Automated System for Computation and Information Processing, abbreviated as OGAS in Russian. Part internet network, part enterprise management system, Glushkov believed OGAS would provide much-needed efficiency to an empire that had outgrown the usefulness of its existing bureaucracy.

OGAS quickly became his life's work. Giving up an outstanding career in mathematics at the age of thirty-two, he switched to cybernetics and was quickly placed in leadership of the USSR's first Institute of Cybernetics, in Kyiv. There, he spent nearly two decades struggling to bring the Soviet Union into the computing future. However, OGAS ended as little more than a file in a dossier at the Central Planning Committee. Glushkov would take his last breaths dictating the unfinished story of his long struggle against Soviet dogmatism. Here's what happened.

1/ Lieberman, Henry (1973). "Soviet Devising a Computer Net for State Planning." [*New York Times*].

Employees at the Institute of Cybernetics' Computing Center
↓

↑
Hard at work: the Institute of Cybernetics' Computing Center

In 1952, eight years before the Institute of Cybernetics would be created with Glushkov as its head, Anatoly Kitov sat reading a book in a secret library within the defense complex SDB-245. The book was *Cybernetics: Or Control and Communication in Animal and Machine*, written by American scientist Norbert Wiener. It was classified.

In those earliest years of cybernetics, the Soviet establishment deemed the field "bourgeois": a purely theoretical pseudoscience "at odds with the principles of Marxism." The Party-controlled media lampooned cybernetics on occasion; it was not taught at the Artillery Engineering Academy in Moscow where Kitov had studied engineering. But Kitov was a voracious learner and eager to apply what he learned. Wiener's ideas made sense to him.

As Kitov moved up the defense sphere, he began drumming up support for cybernetics. In 1954, it paid off.

Kitov was appointed the head of the Computing Center at the Ministry of Defense. Two years later, he published the first computer science textbook in the Soviet Union, *Digital Computing Machines*, the book that prompted Glushkov to switch from mathematics to cybernetics. (In 1962, Pergamon Press published an English translation of a second Kitov book for students in the U.S.)

Kitov recalled these years as a period of rapid reversal in the Party line: it was "the spark that ignited the flame."[2] By 1959, Kitov had compiled a full report on what an automated management system for the Soviet military could look like.

Yet, for all his tenacity (or perhaps, because of it), he lacked diplomacy. The commission convened for reviewing the report was headed by Field Marshall Konstantin Rokossovskiy, one of the highest ranking military officials at the time, and consisted solely of military personnel. Before this audience, Kitov presented a report that included bald criticism of the Ministry of Defense, accusing it of being slow and calling for a total overhaul. Furthermore, his proposed automated management system would curtail the power of the military elite.

The committee's retribution was swift. They did not merely reject Kitov's proposal; they removed him from his prestigious post and stripped him of his Party card. Impressively, Kitov escaped having his entire professional career destroyed by this episode. He continued quietly advocating for cybernetics and helped develop some of the nation's first small-scale automatic management systems in the 1960s. His chance to lead a large-scale automation project, however, was gone.

When Glushkov took up the torch again three years later, the younger scientist seemed off to a promising start. A talented mathematician, his name already held weight in scientific circles for his contributions to solving Gilbert's fifth problem. Friends and colleagues knew him for being charismatic, determined, and having a gift for inspiring others when he saw a worthwhile project.

His early career put him in touch with a wide circle of influential scientists and officials, taking him from Moscow to the Urals and then 2,000 kilometers (1,243 miles) southwest to the industrial cities of Novocherkassy, Russia, and Sverdlovsk, Russia (now Yekaterinburg). Then in 1956, a year after solving one of the hardest mathematical problems in the world, Glushkov decided to switch gears completely. He turned his attention from algebraics to cybernetics — the field of study that would later produce the internet. That same year, Glushkov moved with his wife to Kyiv to take a position as vice-president of the Academy of Sciences of the Ukrainian Soviet Socialist Republic.

It was in Kyiv that he became associated with Kitov. Despite knowing that

2/ Vyacheslav Dolgov, *Анатолий Иванович Китов – пионер кибернетики, информатики и автоматизированных систем управления* [*Anatoly Ivanovich Kitov: A Pioneer of Cybernetics, Computer Science and Automated Control Systems*], (Moscow: КОС.ИНФ, 2010).

the military scientist had fallen out of favor with the Party, Glushkov was hooked by the idea of a state-wide information management system. The two men began working together to apply Kitov's principles to management of the entire Soviet economy.

For several years, Glushkov devoted himself to cybernetics, developing plans for OGAS in private while inspiring other scientists with the broad possibilities presented by computer technology. During this time, Glushkov also became a member of the Communist Party. In 1962, an influential friend at the Academy made Glushkov an introduction that should have sealed the success of his beloved OGAS.

Alexei Kosygin, the First Deputy Chairman of the Council of Ministers, understood the need for modernizing and reforming the Soviet economy. In his later years as the Chairman of the Soviet government, he would enact economic reforms credited for "the golden era" of the Soviet economy. When he met Glushkov, he had just been reinstated as a member of the Politburo. The official was impressed with him and asked him to head the newly created Institute of Cybernetics, with OGAS as his pet project. Glushkov had the green light.

In the USSR, where the Party leadership had the final say, having a man on the inside with some influence to finagle financial and administrative issues would often make or break a project. Kosygin was that man.

Glushkov's new title gave him authority that he'd lacked in his position as head of an ill-funded experimental science department. His command allowed him to monitor the performance of the chairman of the State Planning Committee, all the ministers and ministries, and he could visit any factory he pleased.

Glushkov also received unexpected support from Minister of Defense Dmitry Ustinov, who rounded up all the defense ministry leaders and issued a clear command: "Do what Glushkov tells you to do."

Glushkov's first years in Kyiv at the Institute of Cybernetics were filled with frenetic energy. In 1963 alone, he visited more than 100 production facilities and organizations in various fields, including mines and collective farms all over the country. Under his supervision, the Institute flourished. A generation of Soviet computer and ICS scientists trained there, spurred on by the idea of an orderly, scientific "Cybertopia."

All the while, Glushkov remained committed to making OGAS a reality.

Over the course of the next ten years, he would visit nearly 1,000 industrial sites to study production and efficiency. In his own words, Glushkov had a more complete picture of the economy than perhaps anyone else in the country. But he needed more than statistics to make OGAS a reality.

The project was, simply put, massive. A network of computing centers was

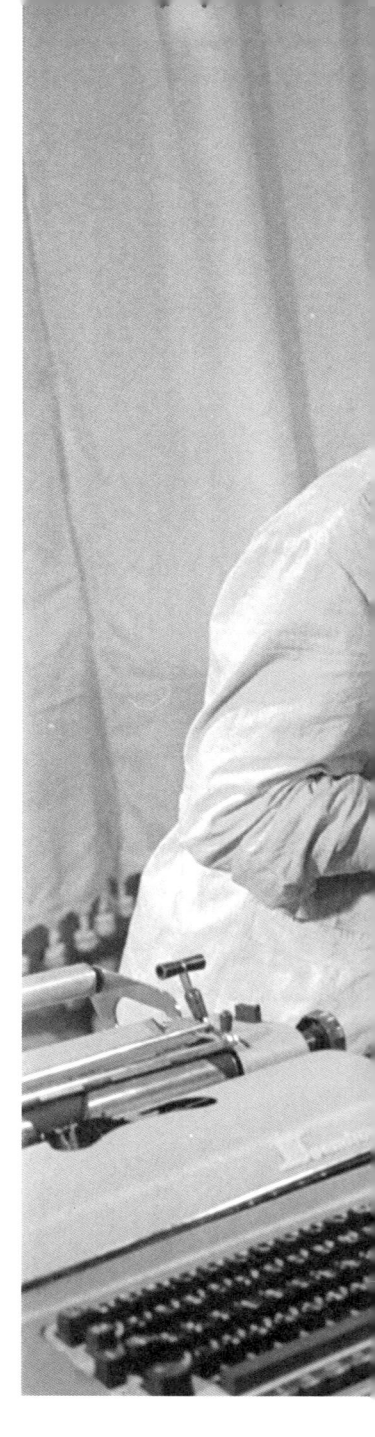

↑
Victor Glushkov, the Director of the Institute of Cybernetics, has a look at a program to be installed on MIR.

needed to service OGAS. Glushkov pushed for building roughly 100 centers in large cities around the USSR which would be connected by broadband communication channels. These would be linked to no less than 20,000 smaller computing centers housed at production facilities.

Glushkov's vision for the early internet system was prophetic. He envisioned broadband communication channels will enable us to connect the computing centers — without channeling devices — so users can copy information from a magnetic tape in Vladivostok to a magnetic tape in Moscow (the distance between these cities was 6,415 km, or 3,990 mi.) without any slowdown. At that time, it would have been a technological breakthrough rivaling the modern internet. Glushkov's plans for OGAS also originally included a cashless payment system, an idea that he was later advised to drop.

In the Party's Politburo, Glushkov's allies were few but powerful. By the early 1960s, it had become clear that the administrative powers of the USSR were reaching their limits. The Soviet model of a centrally planned economy was ill-equipped to handle the economic and population growth of the preceding years. Some of the Party's more pragmatic officials, including Kosygin, predicted that the USSR was on its way to becoming an administrative mess.

As the size and complexity of the bureaucracy expanded, so did the odds of human error. In 1959, a census miscalculation led the population growth to be projected at four million people less than the actual number. When the error was discovered three years later, the state found itself faced with production output planned for a much smaller population. For Glushkov and his allies, the census fiasco was a clear sign that change was needed. OGAS was the solution.

As months ticked on, however, Glushkov would soon find that necessity alone would not be enough to spur the obstinate arms of Party politics to action. The first obstacle in his way was budget. OGAS required 20 billion rubles ($18 billion USD in 1964, according to the official Soviet exchange rate) and fifteen years — an investment rivaled only by the space and nuclear power industries. Glushkov warned an increasingly impatient Kosygin that OGAS could not be accomplished with fewer resources.

The project touched every aspect of life in the Soviet Union, essentially doing what no one, in or out of the Soviet Union, had ever managed. He was adamant that the undertaking would generate a profit. According to his calculations, it would start paying for itself within five years and make a whopping 100 billion rubles over the course of fifteen years. But he needed resources upfront.

The obstacle that would ultimately prove lethal to OGAS, however, had nothing to do with the budget.

In the USSR, scientists working in the defense industry were part of a privileged class. Glushkov was accustomed to dealing with highly ambitious people. Most of them were talented, passionate intellectuals who worked around the clock. Likely, this skewed perception of the Soviet Union as a whole.

The average worker received a fixed salary that was not always enough to make ends meet. Maximizing productivity brought them no gains, so instead, workers survived by subverting or cheating the system with remarkable vigor and ingenuity.

For instance, say a new assembly line was installed at a steel mill. As soon as a new part, still hot, touched the conveyor belt, a heat sensor would indicate that a unit had been manufactured. A worker looking to boost his or her efficiency would simply hold a lit cigarette next to the sensor and have it register a non-existent part. Managers at all levels would take these inflated figures and tweak them a bit further to reach whatever quota had been set by the Politburo.

These strategies were excellent for workers and mid-level bureaucrats to negotiate their lives in a system based on arbitrary production quotas. They weren't excellent for building sustainable industrial power.

An automated system would have exposed the whole system of falsified data. Inflated production numbers, disregard for quality control standards, black market trading — OGAS would have meant game over for the whole lot. It also would have weakened the power of the Politburo. If implemented, this new system would enable users to receive accurate information instantly from anywhere in the country, bypassing the slew of bureaucratic institutions and, ultimately, eliminating the need for them.

Andrei Kirilenko, Communist Party Central Committee Secretary, summed up the situation nicely to Glushkov: "Why do we need production management? I come to a plant, make a speech in front of the workers, and the plant increases its efficiency by five percent! It's much better than your two percent!"[3]

From 1963 until 1970, OGAS bounced from committee to committee, disassembled and reassembled each time by a different minister unhappy with its implications for his bureau. In the mid-1960s, the Cold War was heating up again, and NASA was hard at work to fulfill President Kennedy's promise to put a man on the moon by 1969. It was easy for the ministers to put off Glushkov and his OGAS. The project stalled.

Then in the late 1960s, Glushkov got his lucky break. News broke that the Americans had already designed several computer networks. American computer scientists had started working on creating the forerunner to today's internet later than their Soviet counterparts, but by 1969, they were already on the verge of launching

3/ Borys Malynovskyi, *Академик В.Глушков: Страницы жизни и творчества* [*Academician Glushkov: Pages of Life and Creativity*], (Kyiv: Наукова думка, 1993).

ARPANET, a network that would connect computers all across America. This finally stirred the Party to action. Their primary motivation, however, was security and intelligence. The economic applications of these "computer networks," if any, were secondary.

Instead of letting Glushkov get to work, the Soviet government decided to form yet another commission to finalize the game plan for OGAS. Glushkov lodged a formal protest: "Don't create a committee — that's all I am asking of you. Experience has shown that commissions subtract brainpower, instead of adding it, and they can stifle any endeavor."[4] Nonetheless, a committee was created, composed of career bureaucrats, including the minister of finance, Vasily Garbuzov: exactly the sort of Soviet decision-maker who had more to lose than to gain from OGAS-enabled reforms. Kosygin was Glushkov's only ally on the committee.

On the morning of October 1, 1970, Glushkov adjusted his signature tortoiseshell glasses and strode into the Politburo. He knew that when he walked out again, the fate of his last ten years of work would be decided.

When Glushkov took his seat in front of the committee, he immediately sensed that something was amiss. Neither Kosygin nor Leonid Brezhnev, individuals whose voices on his side could be decisive, were there. Brezhnev was in Baku for the 50th anniversary of Soviet leadership in Azerbaijan; Kosygin was in Egypt attending the president's funeral, but neither had sent word to him. Glushkov did not comment on their absences in his memoir, but it could be guessed that his vision for economic planning was simply not among their priorities. The proceedings began.

It was a short meeting; likely, the fate of OGAS had already been determined. The committee was opened and each speaker took their turn. Glushkov recalled later that when Garbuzov took the floor, he began by describing a machine he had observed on a poultry farm. This "computer" increased egg production by way of well-timed lights and music when chickens laid an egg. "At this

4/ Borys Malynovskyi, *История вычислительной техники в лицах* [*Pioneers of Soviet Computing*], (Київ: фирма "КИ,Т", 1995).

↓
The Bulat computer at the Institute of Cybernetics

5/ Malynovskyi, *История вычислительной техники в лицах* [*Pioneers of Soviet Computing*].

point," recalled Glushkov, "he declared that now all poultry farms in the Soviet Union need to be automated, and only then could we begin to think about such stupid things as the general governmental system. I laughed again and thought: 'All right, whatever.'"[5]

Garbuzov was chided for the comment, but OGAS was defeated nonetheless. The assembled members and Glushkov were told that the government, regrettably, lacked the resources to put the whole system into place and offered a counterproposal: a simplified version of OGAS would be developed as a government network of computing centers. Without any industrial or management control, it rendered Glushkov's idea "a hardware solution without any appropriate software support." Convinced of the need for a scalable alternative to the current production system, Glushkov warned the committee that if the proposal wasn't accepted now, the Soviet system would face dire problems by the end of the decade.

The committee members allowed him to finish, then voted to accept the counterproposal. Once again, ideology and red tape trumped practical concerns.

In the years that followed, cybernetics again lost favor among the Soviet elite. In 1971, news reached America of OGAS (ironically, only after it had been shaved down to bare-bones). Shortly after, according to the memoirs of Borys Malynovskyi, an article appeared in the *Washington Post* entitled, "Punch Cards Control the Kremlin." In it, Victor Zorza, a former Soviet citizen who had emigrated to the U.S., claimed that Glushkov sought to replace the Soviet Union's leaders with computers.

If the US government hoped the article would sabotage the project, it worked. The scenario Zorza painted resonated with the minister of finance, who informed Kosygin that Glushkov was using OGAS to undermine how the Council of Ministers and he personally managed the economy. Kosygin quickly dropped his support of OGAS and began actively opposing it, instead. A second piece by Zorza, this time in the *Guardian*, argued that OGAS was a KGB-run project aimed at spying on the country's citizens.

Glushkov's colleagues, particularly economists, began publishing articles claiming that the Americans had already lost interest in computers. Widely, cybernetics was tarred as a passing fad. Glushkov was instructed to abandon OGAS and focus his energy on more localized projects, like the automated management system at the Lviv Radio Factory.

Undeterred, Glushkov would continue advocating for OGAS for the rest of the 1970s, even as "thaw"-era freedoms were eroded. On the heels of the 26th Congress of the Communist Party of the Soviet Union in 1981, Glushkov launched a final attempt to realize OGAS. His plea "For the Whole Country", was published in *Pravda* (Truth), the main Soviet newspaper.

↑↓
The State Planning Commission's
Computing Center

094

With a circulation of around 12 million, among them Party members and academic institutions who were obliged to subscribe by the government, *Pravda* was good real estate. The piece secured the Party's approval, yet once again, it yielded no practical results.

Alongside Glushkov's long battle with Party politics, he had also been fighting a serious illness for some time. Physically spent, he slipped into a coma at an intensive care unit towards the end of 1981. He awoke after the new year long enough to dictate his memoirs to his daughter, telling her, "I want to make the best out of my last few days on Earth."

On the very same day he finished recounting his battle for OGAS, Ustinov sent his assistant to pay his final respects to Glushkov and asked if the minister could help him in any way. Recalling his early ally's instructions to the staff of the Institute of Cybernetics in 1962, Glushkov exclaimed, "Have him send a tank!" Exhausted by his endless battle with bureaucracy and saboteurs, Victor Glushkov died less than three weeks later, on January 30, 1982.

↑
The head engineer at the Institute of Cybernetics manning the control panel of ES-1060

→
Employees from the Institute of Mathematics standing by a diagnostic machine

chapter 6

↑
A portable black-and-white Yunost-406 TV, manufactured beginning in 1987.
These were often used as home computer monitors.

The USSR and the Rise of Personal Computers

In the Soviet Union, as elsewhere, the first early adopters of the personal computer were radio enthusiasts. A do-it-yourself spirit had characterized the early days of radio. Now, computing filled the vacuum for earlier radio hobbyists. In the United States, the Altair-8800 — the first home computer in the world — and Apple I were both built by amateurs. In Ukraine, the names to remember were the Micro-80, the Radio-86RK, and the best (and the most underrated), the Specialist.

As with many great inventions, the first home computer came into being by accident. In 1978, a computer engineer named Sergey Popov received an unplanned delivery at his office: a parcel from Crystal, a Kyiv-based microelectronics center, containing two microprocessors K580IK80 and K580IK55 — clones of the Intel 8080 CPU. Popov worked at the Moscow Institute of Electronic Machine Building (MIEM). The package was intended for the Moscow Institute of Electronic Control Machines (MIEYM). Popov didn't send it back.

Even though he and his fellow researchers were part of a large Moscow institute, it didn't feel like it when it came to allocations. Popov would later recall that it felt like they were running on empty, their requests for materials frequently ignored. So when the two CPUs appeared on his desk, he quietly tucked them away. A year later, the prototype for the first Soviet home computer was born. It would come to be known as the Micro-80.

The development team's journey to produce it followed a now familiar narrative. Passionate engineers, few resources, and ambiguous government support. At first, the prototype lacked even the essential microchips for read-only memory (ROM); the monitor had to be loaded from punched paper tape. Every morning, the developers would punch in 50 KB of program code manually for the paper tape punch.

Once they finally laid their hands on the ROM chips, they realized they had nothing with which to erase data from it. One of the engineers discovered that a tanning lamp could get the job done. "The lamp would overheat, its spectrum a better fit for people than for microchips," Popov later recalled. "We'd place ROM chips half a meter or so from the lamp, and it'd take half an hour to erase the data. All of us walked around red as beets in the middle of the winter."

Later, salvation came to the sunburned team in the form of a men's facial tanner, a novelty found in a Moscow department store. Despite all the limitations, the Micro-80 prototype was finalized in 1980. It was high time to show it off.

Over the next two years, the MIEM developers courted the institutions that might be interested in the new type of computer. The biggest fish to catch was Nikolai Gorshkov, Deputy Minister of the Ministry of Radio Technology. In the world of Soviet computing, Gorshkov was a big whig. His ministry oversaw the manufacturing of all computing equipment in the USSR and if he could be persuaded to see the value of MIEM, the developers would have a foot in the door of mass production. The meeting didn't go well.

In 1980, twenty years after Glushkov tried to persuade scoffing bureaucrats that the internet was the future, Popov's team met the same resistance about the personal computer.

"I will always remember what he told us that day," Popov later recalled. "'Stop messing around, guys,' he said. 'You just cannot have a personal computer. A personal car, a personal pension, a personal country house — no problem. But a computer? Do you even know what that means? That's three hundred meters of space, twenty-five technicians, and thirty liters of fuel every month!' No one had any idea what that was all about." It was obvious that the support they were looking for would not come from the ministry.

Eventually, Popov decided to sidestep the institutes and ministries all together. He went to *Radio*, a DIY tech magazine, with an article about the Micro-80 miracle computer that could be assembled at home. The plan was to drum up interest directly from potential users. Slowly, personal computers started to gain traction. *Radio*'s print run — almost 900,000 copies — played a big role in that.

In just one year, MIEM had found a way to shrink a computer from the size of a kitchen appliance to the size of an ordinary television set. In fact a television is exactly what was used for the monitors of these early computers. It wasn't a replacement for the computers that powered Soviet nuclear facilities. It was the gateway for an entirely new type of user.

In Ukraine, as in the entire Soviet Union,, personal computing got its wings in street markets and apartment kitchens, assembled by the tech-curious with painstaking effort. Tech hobbyists and engineers would build the skeleton of the Micro-80 system and then complete it with a television monitor and microchips. The Micro-80 had an impressive 200 microchips (the first Apple computer, released in 1976, had just 60). Hobbyists would hunt down microchips from secondhand markets and old machinery, since microchips were not yet being manufactured for consumers.

The Micro-80 kept its singular status for five years, until a vocational school teacher showed up Popov with a new model of a PC built specifically for young people. Anatoliy Volkov taught secondary students in the Ukrainian city of Dniprodzerzhynsk (now Kamianske), and it was likely his work with teens that inspired the project. His idea was to invent a home computer simple enough to be constructed from a minimal number of parts and then sell it in department stores as a construction set for teenagers.

Invent it he did, and soon, the *Trudovaya Smena* ("Shift Change" or "Changing Labor") newspaper was reporting on Volkov's "brand-new kind of construction set" for young

098

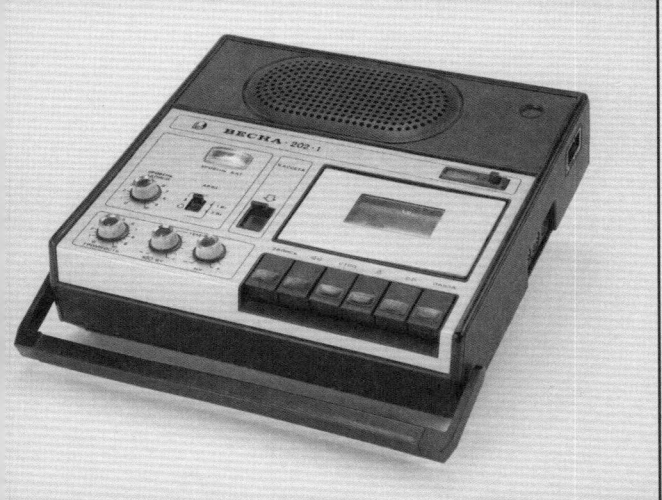

Vesna 202 cassette tape player, often used as a storage device for home computers. Iskra Electric Building Plant in the city of Zaporizhzhia started to manufacture them in 1977, churning out a total of 2 million tape recorders. The cover of *Modelist-Konstruktor* issue that included the Specialist-85 PC review. The text on the monitor read: "The Specialist personal computer built by A. Volkov, teacher at the Dniprodzerzhynsk vocational school. Your personal assistant in studies and manufacturing, household and leisure. More details inside."

It's curious that in the 1980s, magazines didn't just inform their readers about computers, but also published code listings. The readers typed up hexadecimal code manually. Magazines were often available only at libraries and couldn't be taken home, so the readers just copied the code listings into their notebooks.

A nonstarter, the Specialist model never took off, even though in many aspects it was superior to its competitor, Radio-86RK. Suspecting tha his brainchild was doomed, Volkov lost his enthusiasm for its further improvement. He no longer cared about getting the media's attention, so *Modelist-Konstruktor* editors had to dispatch their colleague to Dniprodzerzhynsk to solicit the materials from Volkov.

← In translation: "Electronic computing machine — with one's own hands!" *Modelist-Konstruktor* 1987–2

← Mikrosha PC, the most popular Radio-86RK clone. It was mass-produced beginning in 1987 at Moscow's Lianozovo Electromechanical Plant

The UMPK-R-32 computer, a Radio-86RK clone with 32 KB RAM, produced by the Mukachivprylad factory in Western Ukraine. It was relatively easy to manufacture that model, as it had only 29 microchips, whereas its predecessor, Micro-80, had almost twice as many.

→ A Radio-86RK DIY-computer's circuit board [photo: Wikipedia]

cybernetics fans. He christened it in Ukrainian "Fakhivets"; or in English, Specialist.

The Specialist-85 computer was smart. It knew the BASIC programming language, could draw, play chess, and even make music. It had another major benefit, too — it was widely available. Hobby groups, youth technology clubs, and students of high schools and vocational colleges could all afford the Specialist-85 processor assembly kit. The design used a ready-made circuit board onto which users would attach nodes that came with the kit. After that, they would simply connect a cassette tape recorder and a home television set, and their computers were ready.

What was it about the Specialist that made it superior to other models on market? Its strongest point was that it could display graphics with 384×256 resolution, while the Radio-86RK users had to settle for black-and-white text mode with 64 characters (all uppercase) in 25 lines. The Specialist worked at lightning speed, too, compared to its sluggish rival.

A user recalls the clunkiness of using Popov's model: "The Radio-86RK's video display adapter slowed the processor down. You had to switch off the adapter whenever you had to do a time-sensitive task. The display would just go black while a program started up. The computer just couldn't display and launch a program at the same time."

The Radio-86RK's main goal was simply to work. Buyers of the Radio were mainly interested in learning how computers worked and in building one by themselves. It was the accomplishment they were after; a first step towards professional computer design. But a movement of at-home hobbyists tinkering away on their own computer designs never happened. State-run enterprises decided to take the Radio into production, setting it on a different course.

"They just cobbled together a few available spare parts that cost nothing, a cheap keyboard, and a plastic case — though sometimes they churned those PCs out without it!" one user remembered. "They knew they'd have no trouble selling them."

The unique economic conditions in the Soviet republics at this time created the perfect circumstances for the low-quality Radio to sell. People had money, and nothing to buy. In the 1970s, a computer was a novelty. Even flimsy ones would go like the cakes. Media hype did its part, too. Despite the glaring shortcomings, the Radio-86RK was taken as a basis for an entire series of clones — Mikrosha PC, Spektr 001, Partner 01.01, and other home computer models, which were always compatible with each other.

For all its benefits, the Specialist as a hobbyist computer somehow never took off. It is not clear whether Volkov searched for companies that might have been interested in manufacturing the construction kits. He, like Popov, pitched his invention to *Radio* magazine to drum up interest, but they turned his article down. Popov, on the other hand, after designing the Radio-86RK, had invented a second model, the Radio-86RK, and *Radio* had just finished running a series of articles about it.

Meanwhile, the Specialist went south — or rather, north, to Siberia. Volkov seemingly had stopped caring about his invention's fate. Suspecting that his brainchild was doomed, Volkov lost his enthusiasm to improve or promote it. *Modelist-Konstruktor*, a magazine that had planned a feature piece on Specialist software, stopped getting responses from Volkov. Eventually, they had to dispatch a colleague to Dniprodzerzhynsk to solicit materials from him.

Without a commercial push in Ukraine, a development center from the Siberian town of Barnaul took over the Specialist. It might sound odd given that the Specialist was built in Ukraine, but interestingly enough, the model proved popular in central Russia and Siberia. The Barnaul team finalized the hardware for the first version and developed software for a new software, similar to BIOS, called *ekran* in Russian ("screen").

Developers from a small private company, SP-580, produced reliable commercial software — an upgraded version of BASIC with graphics support, a decent chess program, and a good *Tetris* clone. Software was probably the Specialist PC's weakest spot. Compared to its foreign rivals — for instance, British-made ZX Spectrum with 23,000 programs, many of them games — the Specialist had an extremely modest portfolio. Only a handful of for-profit and hobby development groups coded software.

A programmer from the SP Computer Club estimates that there were not more than 20 active Specialist users in Leningrad. "Software development was a domain for big-city geeks who'd get together and collaborate as a team," he says. "Computer enthusiasts living in provincial towns felt shut out from that movement and rarely moved beyond the regular users' role."

The ADD Group from Dnipropetrovsk (now Dnipro) developed fairly good software for the Specialist, too, including a popular adaptation of Broderbund's 1983 hit *Lode Runner*. It's anybody's guess, though, how much money they made coding and selling software in the USSR in the times of mass piracy, when their competitors just copied and sold programs designed by somebody else.

They got as far as establishing a user's association before the brand-new Orion-128 PC swept the Specialist away in 1991. With both Cyrillic and Latin display capabilities and three graphics modes, the Orion was more advanced and more powerful than the Specialist, though it built on its predecessor's architectural and technical designs.

The Siberian development center shut down in 1992. But the Specialist remained a sought-after model by a certain kind of user. Other development centers took over the project and started to produce circuit boards for homebuilt computers. Later on, a Chernivtsi factory in Western Ukraine also began producing the PC.

THE LIK HOME COMPUTER

A mass-produced Specialist clone. Equipped with a KP580BM80A processor (Intel 8080 clone), 48 KB RAM, and 2-12 KB ROM, in 1991, it cost between $225 and $310. A membrane keyboard with painted letters was reliable, but not very user-friendly, as it didn't give any tactile feedback. It had a Cyrillic keyboard layout, JCUKEN or ЙЦУКЕН, instead of the now popular QWERTY. This was a typical layout for most Soviet computers.

SVEMA MK-60 CASSETTE

Mass-produced and cheap, the compact cassette — commonly called the cassette tape — was probably the most popular data storage device for home computers both in the West and in the USSR. Some of the commercial computers, such as Iskra-1256, produced in the Soviet Union from 1979 to 1988, used it to store information, as well.

Depending on the read-write speed, the compact cassette could have different storage volumes. ZX Spectrum-compatible PCs — one of the most popular Soviet home computers in the late 1980s — worked at a speed of 1,365 KB per second, which meant that a 60-minute audio cassette tape could store around 600 KB of information.

Sixty-minute MK-60 cassette tapes were the most common in the Soviet Union. They were produced at the large Svema factory in the Ukrainian town of Shostka. This enterprise used to be the major photographic film manufacturer in Europe but, like its Western counterparts Kodak and Polaroid, suffered from the impact of digital photography on film. Svema went bankrupt in 2004.

chapter 7

↑
A limited edition of experimental handheld electronic games — clones of the Western models — manufactured in Vinnytsia

→
This photo shows the Motorcycle Racing machine tested by Halyna Taraskina, engineer at the Kyiv-based Elektronmash plant. Courtesy of A. Bormotov. Kyiv, June 1975

Ukraine at Play

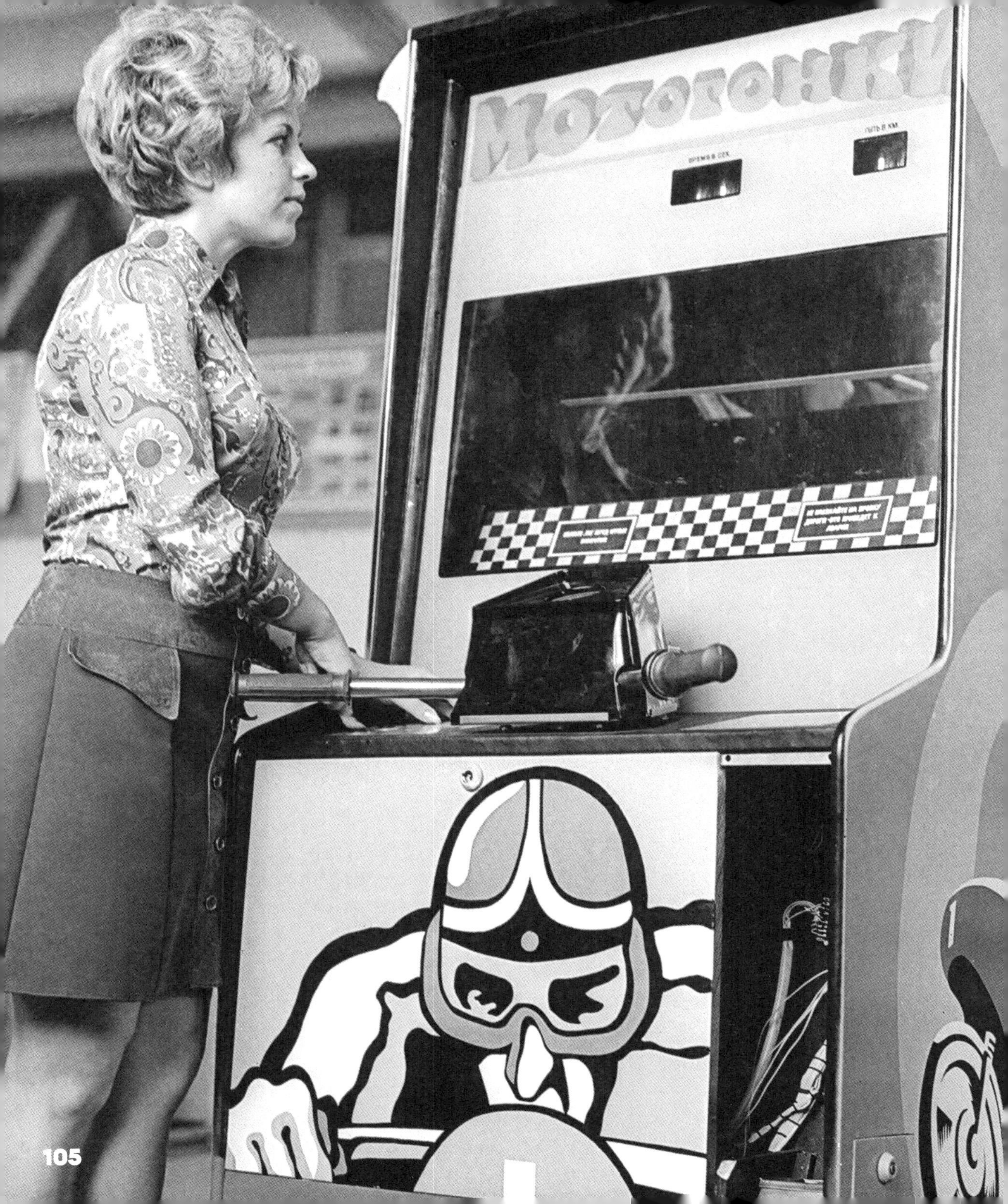
МОТОГОНКИ
ВРЕМЯ В СЕК.
ПУТЬ В КМ

In 1971, citizens and developers from eleven countries flocked to Moscow for the Soviet Union's first International Amusement Trade Show and Arcade Games Expo. Organized by the Ministry of Culture, the open-air event brought more than a dozen of the biggest rides and games manufacturers from the United States, Japan, and Europe. Exciting new fairground attractions like bumper cars and arcade games were debuted for an eager crowd. The exhibition marked the dawn of the gaming industry in the USSR, led by the Soyuzattraktsion company, the first arcade machine supplier in the Soviet Union.

Soyuzattraktsion was tasked with the responsibility of diversifying the leisure of Soviet citizens in public parks. Arcade games were among their specialties. From 1970 until the collapse of the USSR in 1991, Soyuzattraktsion made the games that a generation of Soviet children were brought up on. Funnily enough, most of them were analogues of games from the "forbidden lands," the West. In the same way that state companies would later copy popular Western PC models in the 1980s and '90s, Soyuzattraktsion's development model was buying a sample of each Western slot machine model and then reproducing its design. To cut costs, engineers removed "unnecessary" accessories, simplified the designs, and used much cheaper materials.

Soyuzattraktsion commissioned the machines from military enterprises that had advanced equipment and highly skilled engineers. One former Soyuzattraktsion employee recalled that when quality testing new games at a military plant, they were led blindfolded through the factory to the room containing their machines.

The company churned out dozens of arcade games: car and motorbike races, sports games, pinball, and even claw machines. *Sea Battle* and *The Sniper*, manufactured between 1973 and 1991, were bestsellers. The machines were installed in parks, movie theater lobbies, cafes, and game parlors. The company also produced some merchandise. Some arcade games like *The Submarine* or *Penalty* gave out small branded badges to the winners. The company even printed its own calendars and gave them as gifts to its business partners.

In the mid 1980s, Ukrainians didn't have much access to personal computers with gaming software, so they turned instead to movie theaters and arcades, which installed very basic arcade games in their lobbies. Engineers from Vinnytsia Terminal Plant decided to turn things up a notch and design an arcade game on par with those being made in the West and in Japan. The team, led by Vadym Gerasimov, was eager to design an original product, not simply produce a bootlegged version of a well-known game. Surprisingly enough, the small town engineers pulled it off, despite the dire economic climate and a lack of industry expertise.

Their experiments, run from 1986 till mid-1988, ultimately produced the TIA-MC-1 — a multi-frame color video game machine that functioned as a universal game platform with an 8-bit KR580IK80A microprocessor. Using this as a basis, the plant could have potentially manufactured an unlimited number of arcade games. In the end, Vinnytsia Terminal Plant managed to produce only a dozen. But in the crumbling Soviet Union that was still no small feat.

This kind of leisure was a source of steady profit for the government. Operating the machines at least ten hours a day would offset the manufacturing and installation costs in as little as one year. At the same time, huge profits had a downside — glaring corruption in the industry. It was hardly possible for the government to track how each machine worked or how much profit each location generated. Managers of cafes, movie theaters, and arcades could (and did) simply send the required amount of rubles at the end of each day and pocket the rest.

The TIA-MC-1 model kick-started an entire lineup of arcade machines installed across the Soviet Union, many of which — *The Snow Queen, The Fisher-Cat, The Kotyhoroshko* — were based on famous fairy tales. The most popular arcade game from the TIA-MC-1 series had a curious name, *The Little Humpbacked Horse* ("Konyok-Gorbunok" in Russian). The game was based on a well-known Russian fairy tale about the heroic adventures of Ivan the Fool and his magical horse.

The machine operated just like any other Soviet model. Users would drop a coin into a slot and watch a short animated intro followed by a game set in a fantasy 8-bit world. The goal was to collect all the prizes: the Firebird, the treasure trove, and the beautiful Tsar-Maiden.

Players controlled characters with a joystick. They moved the stick to go left or right, jump or lay down on the ground, or pressed the button to punch. When a character reached the right-hand limit of the screen, the game moved on to a new level. There was a progress indicator at the top of the screen. The time left, the number of lives, the total score, and the level were all shown at the bottom. Whenever a user goofed up, sending their character into a puddle, against a rock, or into a fireball, they lost one life.

Old-school Soviet gamers recall that *The Little Humpbacked Horse* was not as easy as its innocent title suggested. A "Game Over" message would also flash once a player's five minutes were up, even if they had one or two lives left. So, the player had either to drop another coin into a slot or let someone else have a turn. In movie theater lobbies, dozens of people would stand in line for the games, all eager to have a go before the movie started.

The Little Humpbacked Horse's gameplay resembled that of *Pitfall!*, a legendary game for the Atari 2600 console, released by Activision in 1982, that helped define the side-scrolling platformer genre. Its graphics, though, were what really pushed the limits. A color picture tube displayed the game at 256×256 resolution, sixteen colors per level. The screen was divided into smaller sections of 8×8 pixels, with the background image made of 1,024 (32×32) tiles.

This breakdown scheme allowed the RAM to hold 256 tiles that occupied only 8 KB in total. Had the display method been different, the graphics of just two game levels would have taken up 64 KB — the entire addressable processor memory.

Given this output, it is a shame — although not surprising — that Soviet gaming enterprises never reached the level of EA Games or other western counterparts. Mykola Shcherbyna, a seasoned software developer and vintage computers collector from Lviv, says that a lack of a regulated software market played a cruel trick on the TIA-MC-1.

"Game parlors opted for arcade machines based on the Atari console, with the popular *River Raid* shooter, or on the ZX Spectrum with an Exolon shoot-'em-up game. Competing with pirated games was next to impossible — it cost a fortune to develop brand-new software, so the locally manufactured arcade machines carried a hefty price tag."

Another problem was that the state-run Vinnytsia Terminal Plant that produced them was also having a bad streak. Eventually, it turned to gambling games to generate more profit. Shcherbyna remembers that the company designed a special MP1-14 platform that served later as a basis for *Gold Luck!* poker game, strip poker, black jack, Russian roulette, and other games like that. In 1993, the software engineering team set up an EXTREMA company that developed a range of platforms and gambling games.

Yet the TIA-MC-1 platform has not been lost to gaming history. Thanks to computer geek Oleksandr Semenov, it still exists in the form of a freely distributed TIA-MC Emulator for Windows with vintage games — the legendary *Little Humpbacked Horse* among them.

Despite low-quality maintenance and neglect, many of the arcade machines survived till the early 2000s. Retrogamers and urban explorers sometimes come across them in abandoned Soviet summer camps and health centers, in boarded up movie theaters, or in the basements of village cafes. Today, vintage arcade machines are considered Soviet-era artifacts and a valuable find for collectors.

HANDHELD ELECTRONIC GAMES

In 1980, the Japanese company Nintendo launched Game & Watch, a brand new gaming product that promised to be a hit. The gadgets, which took off in the West, soon enough caught the eye of Soviet engineers. As early as 1984, Game & Watch clones were being produced at several Soviet enterprises, three of them based in Ukraine.

Every gadget had an Elektronika label that belonged to the Ministry of the Electronics Industry and initially cost 25 rubles — one-fifth of an average engineer's monthly salary. They didn't get much cheaper with time. Up until the late 1980s, electronic games were hard to find, readily available only in Moscow. The most popular game was a clone of Game & Watch's *Mickey Egg*, in which a wolf had to catch chicken eggs sliding down four conveyor belts.

↑ An Elektronika handheld electronic game, one of numerous Soviet clones of Nintendo's Game & Watch

By the late 1980s, enterprises across the Soviet Union had produced over several million gadgets, some of them high-quality, the others mediocre. In the early 1990s, demand collapsed, and their manufacture gradually came to a halt.

The Zhovten plant in the Ukrainian town of Vinnytsia tried to rise above its competitors by producing a few "original" models. To what extent they were original is a big question. For example, *Winnie the Pooh* and *Space Adventures* are clones of Nintendo's *Donkey Kong Jr.* and Tomytronic's *3-D Planet Zeon* respectively. Today, these survive only as rare gems of private collections.

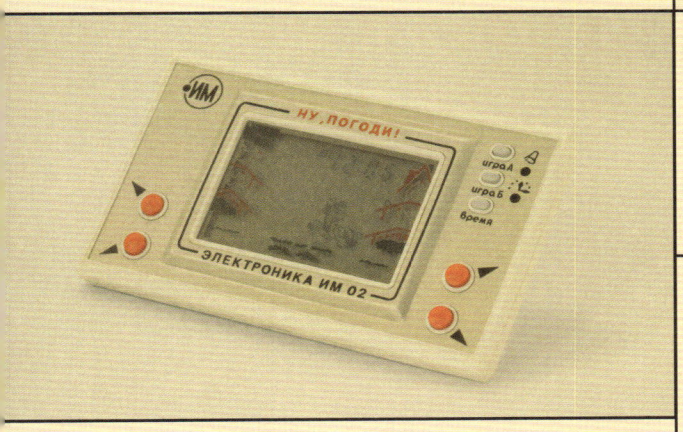

↑ *Well, You Just Wait!*, a handheld electronic game, one of the numerous Soviet clones of Nintendo's Game & Watch. The American Mickey Mouse was replaced by a wolf — a character from the eponymous Soviet cartoon.

→ The Gorodki arcade machine based on TIA-MC-1 platform, 1991

↑ Soyuzattraktsion corporate logo

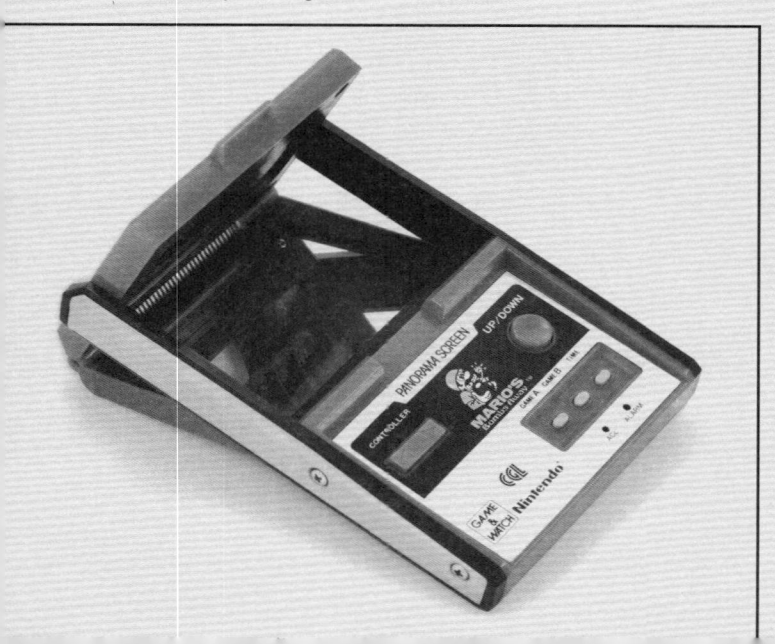

← A limited edition of an experimental handheld electronic game — an imitation of Western models — manufactured in Vinnytsia, Ukraine

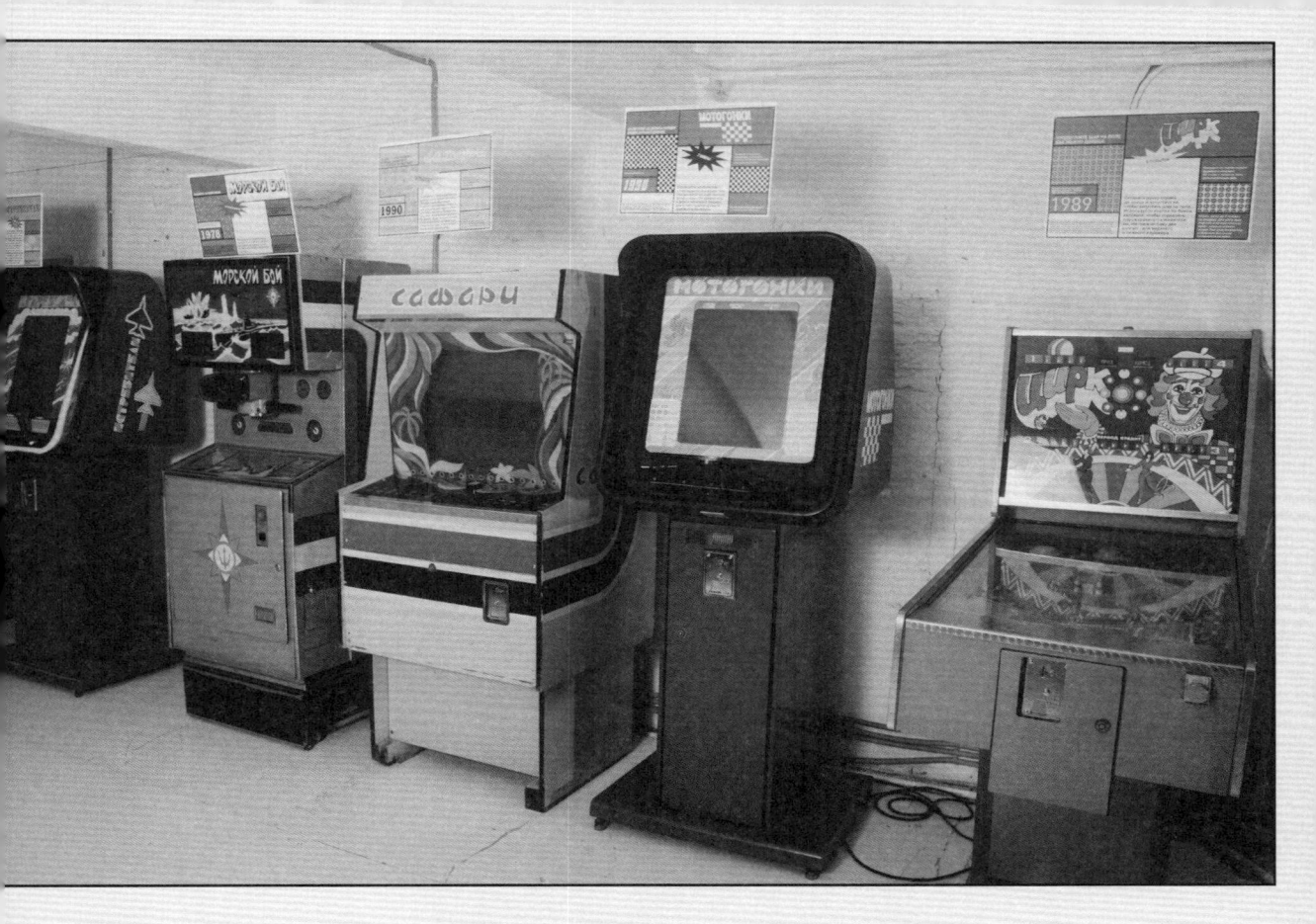

Soviet arcade machines (left to right): Fighter, Sea Battle, Safari, Motorcycle Racing, and Circus.

Back in the 1980s and early 1990s, school kids queued up in droves to try their luck to try their hand at Gorodki arcade machine, based on an ancient sport of the same name. Players had to throw the stick to knock out 15 wooden figures moving from right to left. Poor black-and-white graphics were compensated for by fascinating gameplay. Relics like the Gorodki, as well as other arcade machines manufactured at the Vinnytsia Terminal Plant, were original creations rather than clones of their foreign counterparts like many machines produced elsewhere. Today, gaming nerds can sometimes come across superstar Gorodki arcade machines in aging small-town cafes, if they're lucky.

chapter 8

↑
The first Lviv PC

An All-Ukrainian PC

At the same time that Anatoliy Volkov was pitching his PC prototype to the media from present-day Kamianske, Volodymyr Puyda was hard at work 1,000 kilometers (621 miles) away in Western Ukraine, developing his own model. Puyda was a brilliant young researcher at Lviv Polytechnic Institute's Special Design Bureau, a restricted institute that worked on prototyping advanced technology. The tech developed there usually had a direct pipeline to the defense sector, and computers were a big part of that. Long-range missiles, the nuclear project — the bulwarks of Soviet defense relied on the bulky Dnepr and models like it.

Small computers, the kind that were advertised in magazines like *Radio*, were for armchair engineers and state-run enterprises. But Puyda foresaw that the microprocessor would become indispensable in the defense industry.

Puyda had heard about Motorola's and Intel's successes developing microprocessors. Unwilling to copy the Western models, he decided to create an original product matching his own vision. That computer became the PC-01 Lviv. Unlike most Soviet personal computers, PC-01 Lviv was not a clone, although its processor was similar to the Intel 8080. Although it would have been far simpler to replicate a Western model, Puyda wanted the innovation that would come along in the process of building something from scratch.

"People often ask me why I didn't just copy foreign computers like Commodore 64, ZX Spectrum, or any other model," he would later say. "Well, I was against all this copying and cloning, since I believed that it killed creativity."

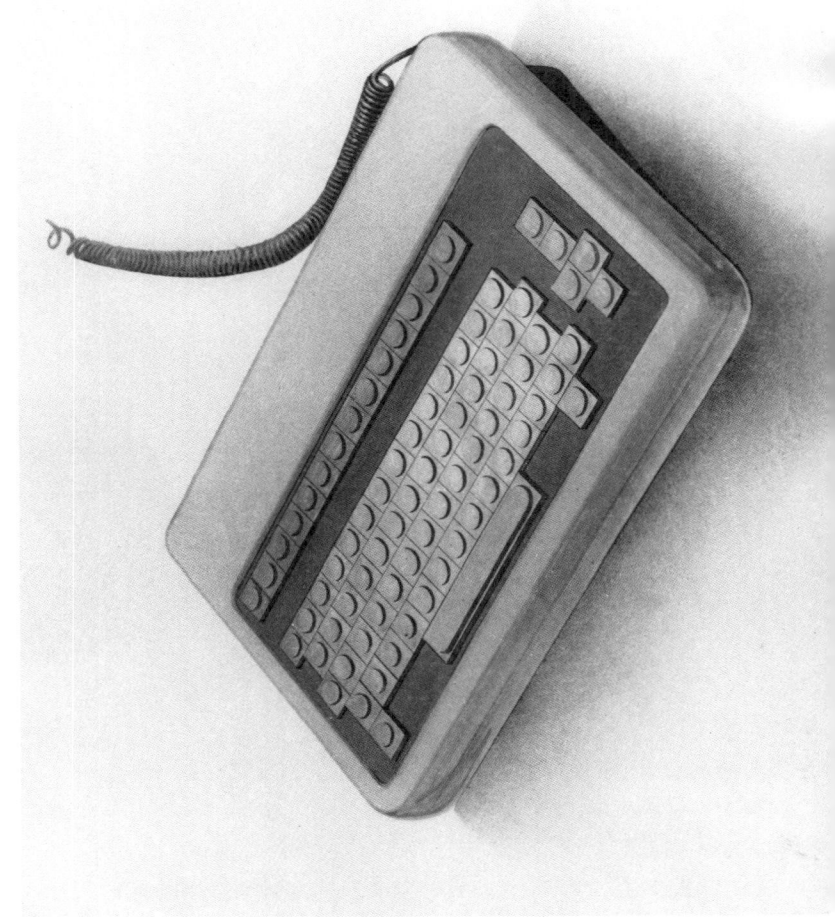

Without access to any of the technical documentation Western scientists had compiled, Puyda's path was far from straight forward. He had to request special permission to stay at the lab

↑
The first mockup of the Lviv PC. In mass production, another prototype was used.

Volodymyr Puyda, the Lviv PC creator, says:
"People often ask me why I didn't just copy foreign computers like Commodore 64, ZX Spectrum, or any other model. Well, I was against all this copying and cloning, since I believed that it killed creativity. It was the wrong path from the technological point of view, too.
In that case, I would've had to reproduce everything down to the smallest construction details, which was next to impossible, since we didn't have access to cutting-edge technologies or manufacturing licenses. It made sense to buy them, of course, but at the same time, we had to develop our own knowledge-based production. I decided to design my PC on the basis of the 580IK80 microprocessor — the Ukrainian analogue of the Intel 8080A, which was widely used elsewhere in the country. Moreover, the 580IK80 was certified for use in the defense industry."

after hours and toiled away at home on the weekends until his design was ready. Puyda was a team of one.

Hardly anyone but him saw the designs until 1985, when he got the chance to present the Lviv at the All-Union Exhibition of Achievements of National Economy (VDNKh) in Kyiv.

The permanent exhibition center showcased the greatest achievements of the Soviet republics. Nicknamed the "Soviet theme park," VDNKh attracted thousands of spectators each year, among them Soviet officials who had the power to push forward projects they liked. The show was a success. Puyda got stellar reviews and was awarded Best in Show for his computer.

Buoyed by his resounding success at the trade show, Puyda boarded his return train and immediately set to work making improvements on the model, right there in the carriage. "On the train ride home," Puyda recalls, "an interesting idea came to me — combining random dynamic access memory hardware with the display and programming a graphic screen. And in synthesizing all the information, including all the alphabetical and numerical symbols, with a special program."

With the VDNKh award, the winds shifted favorably for Puyda and the Lviv PC. The rector of the Lviv Polytechnic Institute, Mykhailo Havryliuk, provided robust, possibly decisive support for mass production. He also deserves credit for convincing state officials to keep the name "Lviv." The multi-national Soviet Union didn't always embrace the idea of one republic taking ownership over an accomplishment, so he had to fight hard to keep the Ukrainian city as its name.

The first commercially-produced Lviv PC was made at Puyda's design bureau and LORTA, a local factory, began mass production in 1987. The first Lviv PCs had a metal frame, but later on, the manufacturers switched to a colored plastic one.

For its time and class, the Lviv PC possessed decent graphic capabilities: 256×256 pixels, pixels, four colors, and a built-in BASIC interpreter with graphic operators, which enabled users to draw color shapes on the screen. Not half bad for an inexpensive PC from the mid-1980s! The computer was equipped with a regular tape recorder and customers received a cassette loaded with Text Editor, Assembler, Reassembler, and other programs.

Despite the fact that LORTA was one of the best factories under the auspices of the USSR's Ministry of Radio Technology, the print circuits and keyboards on the Lviv PC weren't exactly of the highest quality, nor the most consistent. A case in point was the stylish but impractical keyboard. Its letters could withstand a tremendous amount of wear, yet the keys were uncomfortably hard to press.

The planned economy was to blame for that. The government hadn't allocated LORTA any funds to purchase the right kind of metal for the little springs. The keyboard layout — JCUKEN instead of QWERTY — was rather odd, too.

Later on, other factories began manufacturing the Lviv, as well. Although the Lviv PC didn't have extensive application software, especially compared to its main competitor, ZX Spectrum clones, it sold pretty well up until the collapse of the USSR. Due to its vast functional capabilities and small size, the Lviv PC became a fixture at research centers and factories. According to various estimates, a total of 80,000–90,000 Lviv PCs were produced, making it one of the most popular Soviet PCs.

The Lviv was quite affordable, relative to other models. It cost 750 rubles, which was about four to five months' salary. By comparison, Soviet clones of IBM PCs cost between six and sixteen times as much. When Information Technology became a part of the national school curriculum in the late 1980s, the Lviv PC took up residence in classrooms across the USSR. LORTA even created a mobile classroom for schools in rural areas, where hardly anyone had heard about computers. Many of today's computer engineers first learned how to program with BASIC and Assembler on the Lviv PC.

With the 1990s on the horizon, plans for developing software at the factory were drafted, along with plans for two new Lviv models — the Mukachevo PC-02 and PC-03. However, none of these ever came to be. Ultimately, the Lviv's success was stymied by the same forces that helped bring the Soviet Union to an end: namely, a sluggish economy.

Curiously, Puyda never copyrighted the Lviv, foregoing the royalties that doing so would have brought. But for a Soviet-era computer scientist, his story has an unusually happy ending. As a reward for his contributions to cybernetics, the Soviet government provided the young computer whiz with a four-room apartment in Lviv, as well as a generous monthly salary and other perks. After the collapse of the Soviet Union, Puyda was part of the vanguard of computer scientists and programmers that led IT development in a newly independent Ukraine. Currently, he continues to work in IT from Lviv, where he also teaches classes at Lviv Polytechnic National University.

→ Professor Volodymyr Puyda working on modernizing the Lviv PC-01 with his colleagues and students at the Lviv Polytechnic University

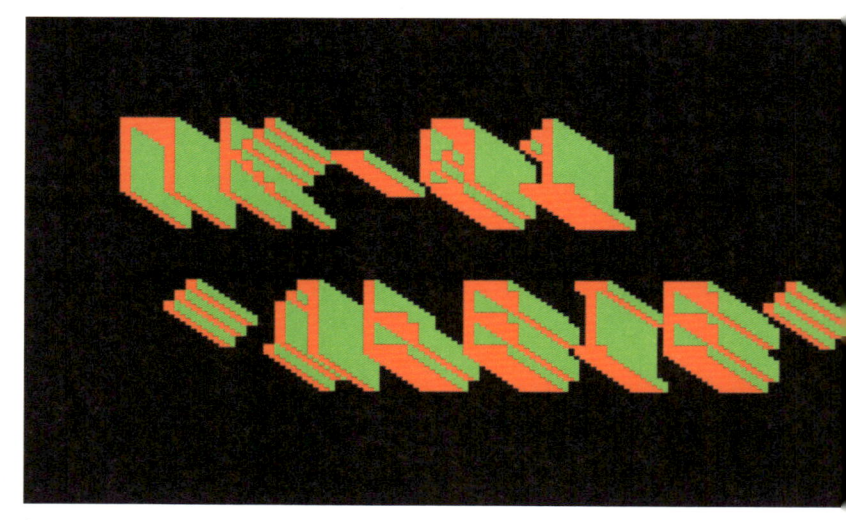

↑ The start screen of the Lviv PC with the name of the computer in Ukrainian

→ A KR580VM80A processor, 2.22 MHz, compatible with Intel 8080, 64 KB RAM (including 16 KB video memory, 16 KB ROM)

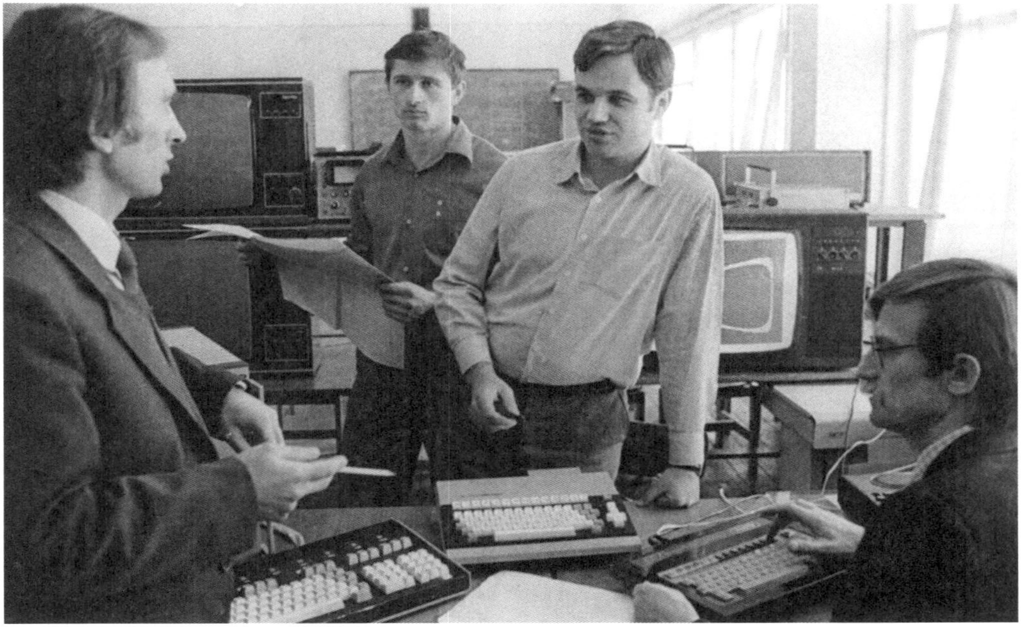

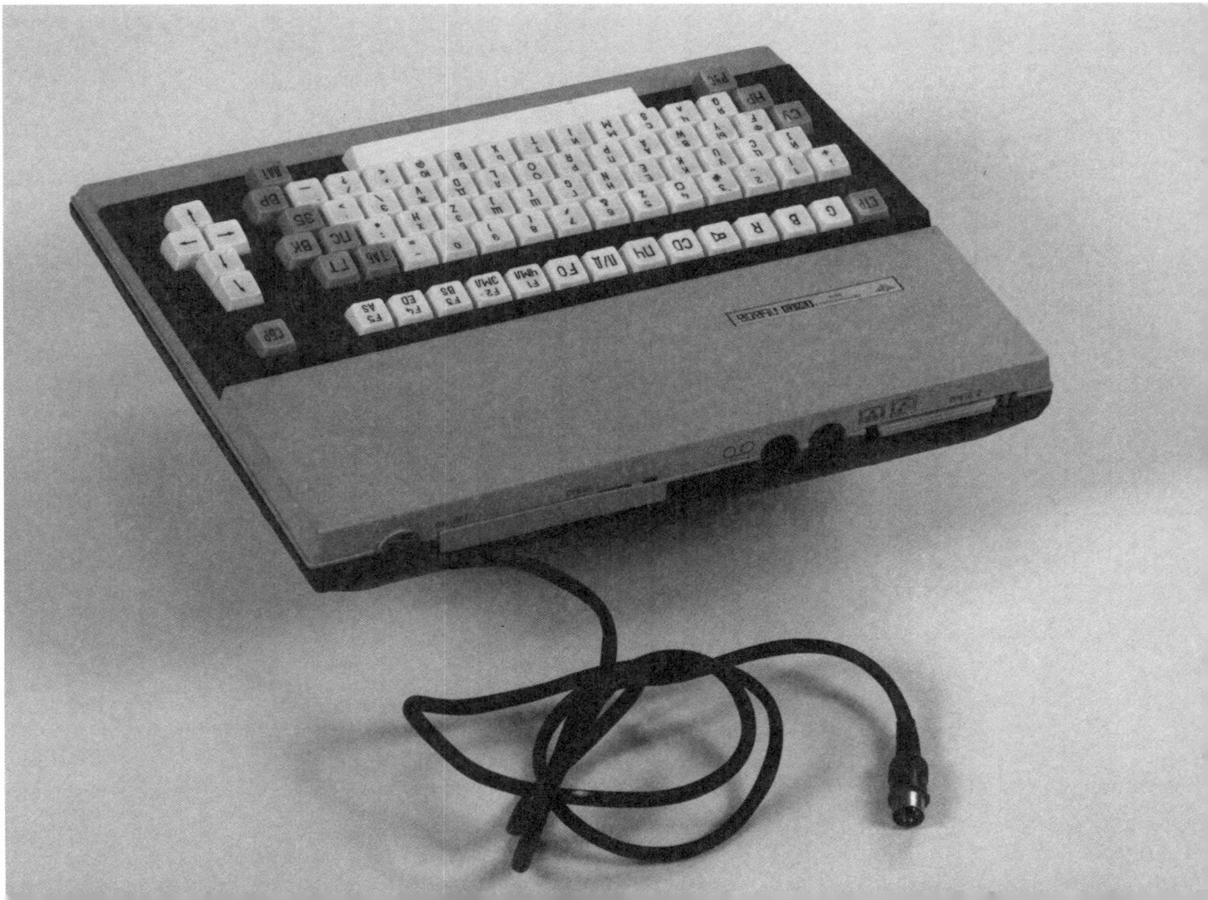

chapter 9

↑
The Parus VI201, a Spectrum-compatible computer produced in the
early 1990s at the Sevastopol Electromechanical Factory in Crimea

ZX Mania

For decades, the public in the U.S. and across the world had eagerly awaited *Time* magazine's announcement of "Man of the Year" every winter. At the close of each year, the magazine's editors deliberated to choose the single individual who had most influenced the course of world events in the past twelve months. Previous choices had included Mahatma Gandhi, Queen Elizabeth II, and Winston Churchill. In 1982, the much-anticipated cover hit newsstands with a surprise: the "Man of the Year" was The Computer.

This was the dawn of the Computer Age. Personal computers came out of the lab and into the living room, with the charge led by British technology company Sinclair. Their affordable ZX Spectrum hit the market in 1982 and quickly gained popularity all over the world. In the U.S., Timex teamed up with Sinclair to manufacture Spectrums. Its market price was $99.95 (about $319 in today's dollars), and it was advertised as the most affordable home computer. Meanwhile in the Soviet Union, tech nerds set to work making Spectrum clones.

From the Caucasus to the Danube, young people were gripped with Spectrum fever. Knockoff Spectrums took over the budding home computer market in the late 1980s, eclipsing Ukrainian and other Soviet-made models like the Lviv and the Specialist by the early 1990s. Thousands of enterprising former Soviet citizens made money by assembling and then selling would-be Spectrums, hawking pirated software on cassettes and disks, and printing user manuals and game reviews.

A big part of the Spectrum's popularity was the vast selection of games it could run. For a few rubles at local markets, students could buy cassettes preloaded with analogues of games like *Pac-Man* and *Knight Lore*. Spectrum favorably differed from other affordable Soviet PCs. It not only had fancy graphics and animation, but users could also connect it to a special sound co-processor and get three-channel stereo sound.

Playing games was much more fun and gave those who made electronic music a wider range of opportunities. In the USSR, Spectrum provided many with a new artistic outlet — more creative programmers would compete against each other to see who could make the best demoscenes.

The machine's simplicity was part of its appeal. Anyone with the money to spare could simply go to the store, buy a Spectrum kit, and assemble it at home. More devoted techies could buy ready-made circuit boards and make their own computers that way. Of course, DIY Soviet Spectrums often had glitches. Sometimes, users simply didn't know how to assemble them properly. At others, the microchips were incompatible or the Soviet-produced parts failed to work. Then there was the problem of power spikes, which could fry the computer's circuit boards in the blink of an eye.

↑
The Timex Sinclair 1000 was the first licensed American clone of ZX81, the predecessor to the ZX Spectrum. Roughly 500,000 were sold within the first six months.

←
The Orel BK-08, ZX Spectrum clone manufactured at the Dnipropetrovsk Machine Building Plant. This computer was equipped not with the Soviet Z80 processor but with the UA880A, produced in the German Democratic Republic.

← The Timex Sinclair 1500 is a modified version of the Timex Sinclair 1000 with its signature tape recorder and collection of games on display at Club8bit, a computer museum in Mariupol, Ukraine.

↓ The Parus VI201, a Spectrum-compatible computer produced in the early 1990s at the Sevastopol Electromechanical Factory in the Autonomous Republic of Crimea, Ukraine

Ikar, a Kharkiv-based company, produced a Spectrum clone by the same name. The PC itself was designed and manufactured by Elektropribor, now known as Khartron, a state-run factory in Kharkiv that designed and produced rockets and space equipment.

For this reason, the metal frame of the Ikar was much more durable than its counterparts

↑
A Delta home computer, produced starting in 1989

←
Robik was a relatively popular, yet mediocre Spectrum clone produced by Rotor, a factory in Cherkasy, Ukraine. It was roughly only 40% compatible with ZX Spectrum because certain changes were made to Russify the interface.

Spectrum was both a mainstream and an underground computer in the late-Soviet and early post-Soviet period. Programmers learned BASIC, played the best foreign and Soviet games, composed music, and made demoscenes on them. No other computer spoke to Ukrainians' hearts as much as Spectrum did. That was, until IBM-compatible PCs appeared in the late 1990s.

chapter 10

↑
The Sevastopol Electromechanical Factory in the
Autonomous Republic of Crimea, Ukraine,
early 1990s

The Life and Death of Soviet PCs

In the first thirty or so years since the first electronic computers were created, numerous competing platforms emerged. Outside of the USSR, Apple II, Tandy TRS-80, and Commodore PET were the big names. It wasn't long, though, until IBM took center stage. In 1981, the company, supported by Intel and Microsoft, produced its own PC, the IBM Personal Computer 5150. The IBM PC-5150 demolished its competition in the West and would come to set a new standard for personal computing in the East.

Soviet factories, including ones in Ukraine, quickly set about cloning IBM PCs. In the mid 1980s, the Kyiv Research Institute of Radio Equipment developed the Neuron I9.66. This PC was compatible with the IBM PC/XT, a second-generation IBM PC that used a new kind of processor.

The computer stood out for its large system unit and numerous storage devices. The Neuron could operate four disk drives that took 740 KB floppies and one MFM hard drive of 5 or 10 MB. (Now, it's hard to imagine why anyone would need so many.) The Neuron's operating memory was much less: 256 KB, which was unexceptional even for that time, or 512 KB. Nevertheless, its designers called it a "professional personal computer," meant to automate measurement tasks.

Each of the computers was equipped with one of two operating systems: Neuron-DOS1, a clone of MS-DOS 3.0, or Neuron DOS2, compatible with CP/M-86. Although the names were similar, the OSs were quite different. CP/M was standard on computers with Intel 8080 and Z80 processors, just like MS-DOS was for the Intel x86 platform.

Soon enough, in the late 1980s, the PC/XT was no longer cutting-edge technology and its hardware became cheaper. A Kyiv factory called Electronmash began manufacturing a new model, the Poisk. It was the first more or less affordable IBM PC-compatible model and eventually acquired cult status. At first, educational institutions were its main buyers. The second model hit store shelves in 1991, right before the Soviet Union fell.

The Neuron I9.66, a Sovietmade PC. The top section (directly under the monitor) was for two disk drives and the hard drive.
↓

Neuron had an I41 system interface (a clone of the 32-bit Intel Multibus), which, according to the manufacturer, "enables multiprocessing configuration and network architecture."
In other words, the Neuron could rightfully call itself a professional personal computer. But was it a high-quality product? By today's standards, the assembly technology used is simply mind-boggling. The pieces of hardware were connected not by a printed circuit board nor even soldered. Instead, all the wires were fastened to special plugs. So, you had to be extra careful when handling the Neuron.

↓
The Elektronika MS6105.02, a common Soviet black-and-white monitor

→
Assembling PC boards for electronic machines at the Computing and Controlling Electronic Machines Plant in Kyiv

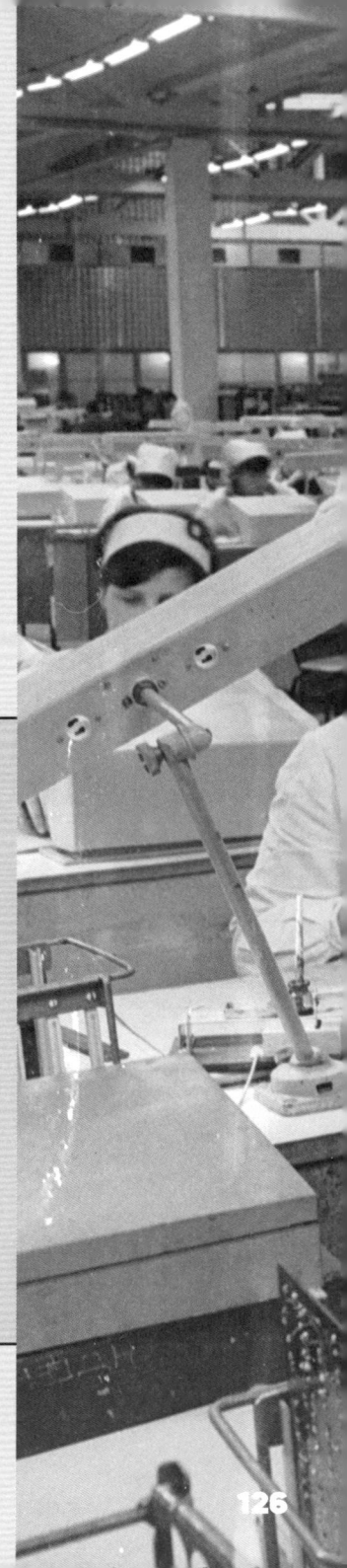

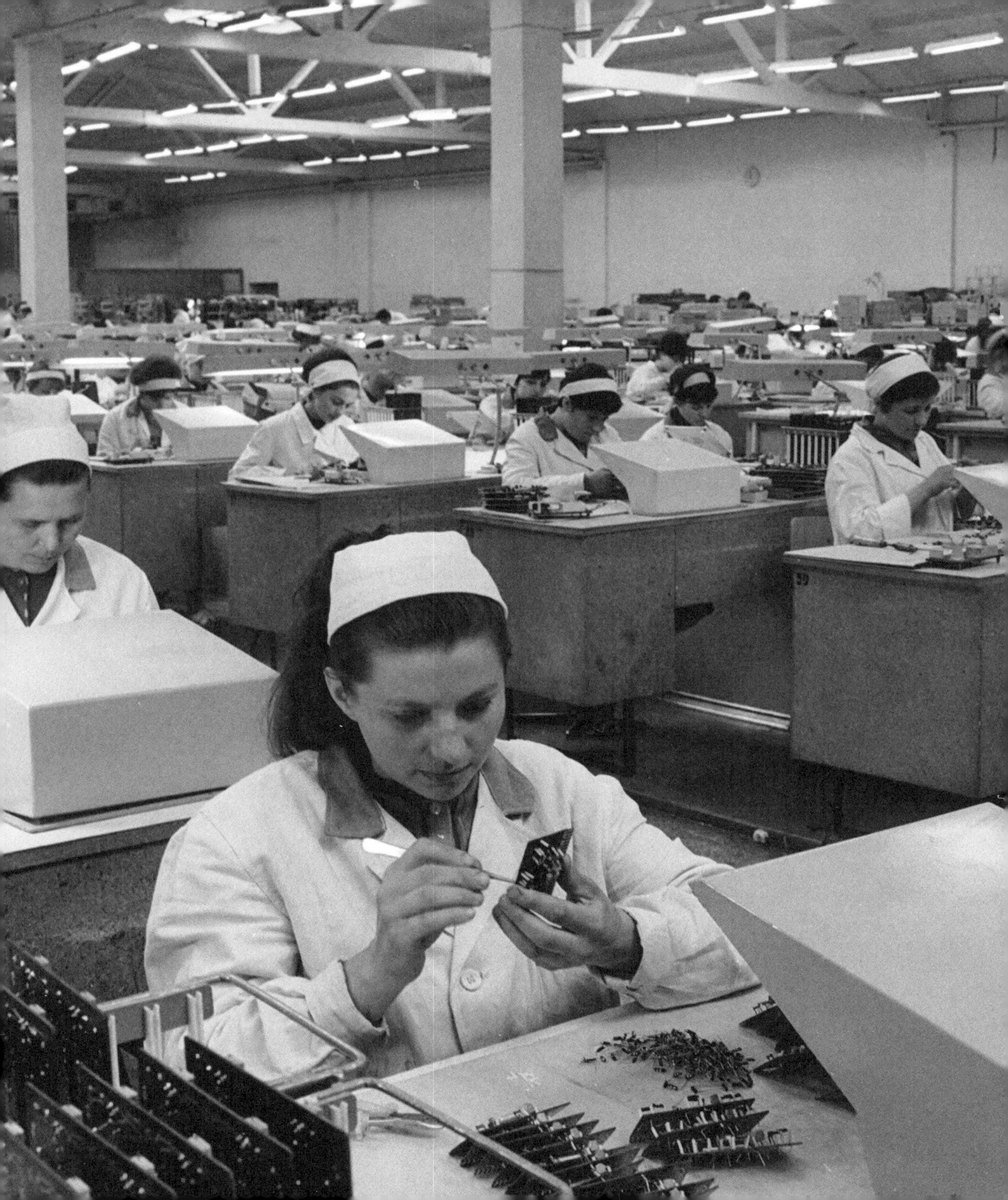

The Poisk, which translates to "the Search" in English, was an all-in-one PC that connected the motherboard and the keyboard in a flat plastic body. The disk drive controller, hard drive controller, RAM expander, and many other boards could be attached to the top of the frame as cartridges, giving the computer the odd look of a modernist building.

Frankly speaking, the Poisk was a cheap computer that paled in comparison to the PC/XT. It was about as bare bones as PCs come, so users had to source additional boards and other parts independently to make their user experience more or less tolerable. And the parts it did come with were poor quality. One can only assume that in the final months of the Soviet Union, the state had more pressing matters than ensuring high-quality microprocessors reach Soviet citizens.

The Poisk was also significantly less efficient than the XT. It lacked Direct Memory Access (DMA), so the cheaply made Soviet processor was overloaded. It could barely even run applications. Depending on the model, the Poisk had 128 KB or 512 KB worth of memory, and 32 KB were allocated for the VGA. DOS took up a great deal of memory, leaving hardly any for games. The situation changed with the addition of a hard drive or disk drive. "Poisk literally came alive when equipped with a disk drive and a hard drive, and it became a much more legitimate computer," said one user.

Poisk computers had a reputation for unreliability and poor user experience. "What kind of memories do I have?" said one former Poisk user. "It was like a nightmare! I had about three or four Poisks. And that wasn't because I was a big fan. Each new one had some sort of glitch. I'd take it to get fixed, and then I'd buy another one while it was getting fixed. The new one started acting up on me, too, even though it had a 512 KB memory card. I traded it for a Poisk 1, which only had 256 KB."

At the peak of its popularity, the Poisk PC and Elektronika MC 1502 faced off for user loyalty. Both were IBM-compatible personal computers of Ukrainian make. Developed at Microprocessor, a Kyiv-based research and production association, the MC 1502 model did not gain as much traction as its rival at first, even though it boasted more advanced features.

The MC 1502 featured a fully functional graphics controller that was compatible with CGA and sixteen colors (the Poisk had only four), 16-KB ROM (vs. 8-KB ROM), a built-in power supply unit (the Poisk used an external one), BASIC written into the ROM (supplied on a separate ROM-cartridge), and a floppy-disk controller. The MC 1502 was also cheaper.

The competition became moot when the Coordinating Committee for Multilateral Export Controls was eliminated in 1994. This committee had prevented Western companies from selling strategic goods to the Eastern Bloc,

→ The Poisk, equipped with a 5¼-inch disk drive, joystick, and various expansion cards

→ Expansion cards

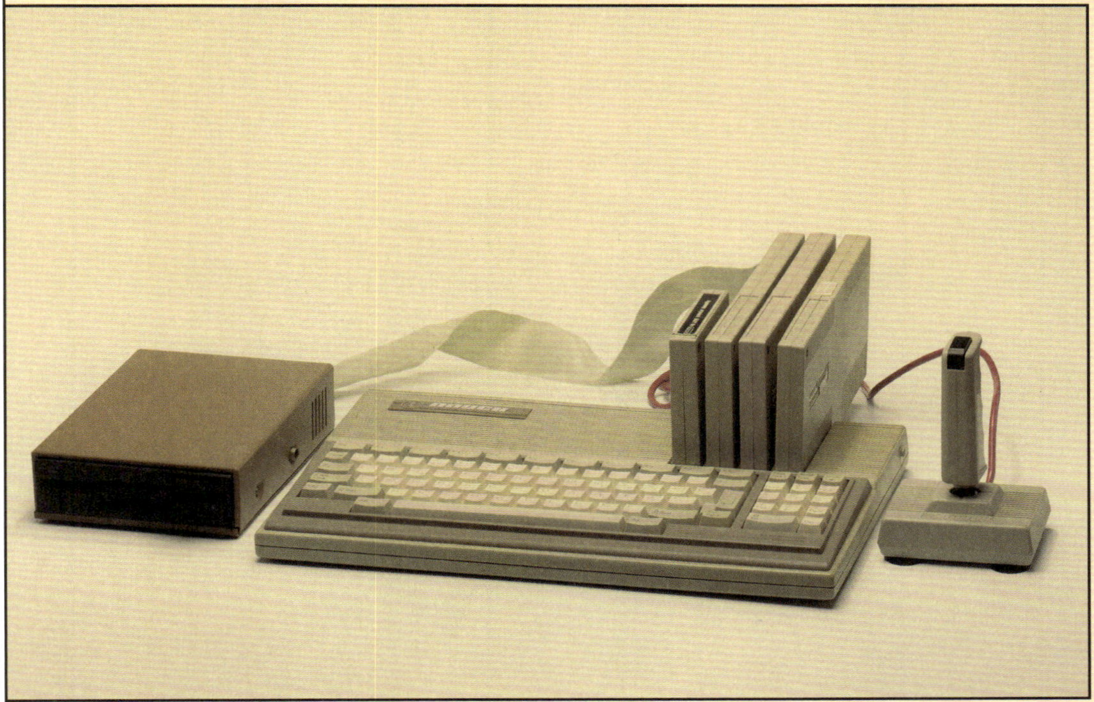

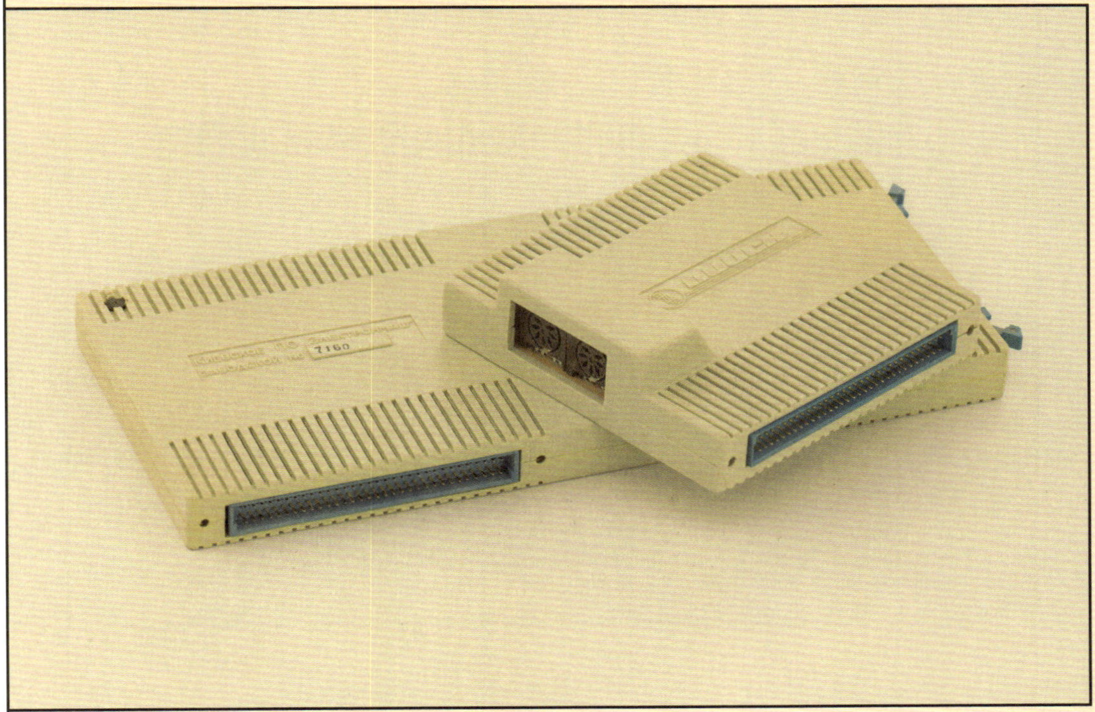

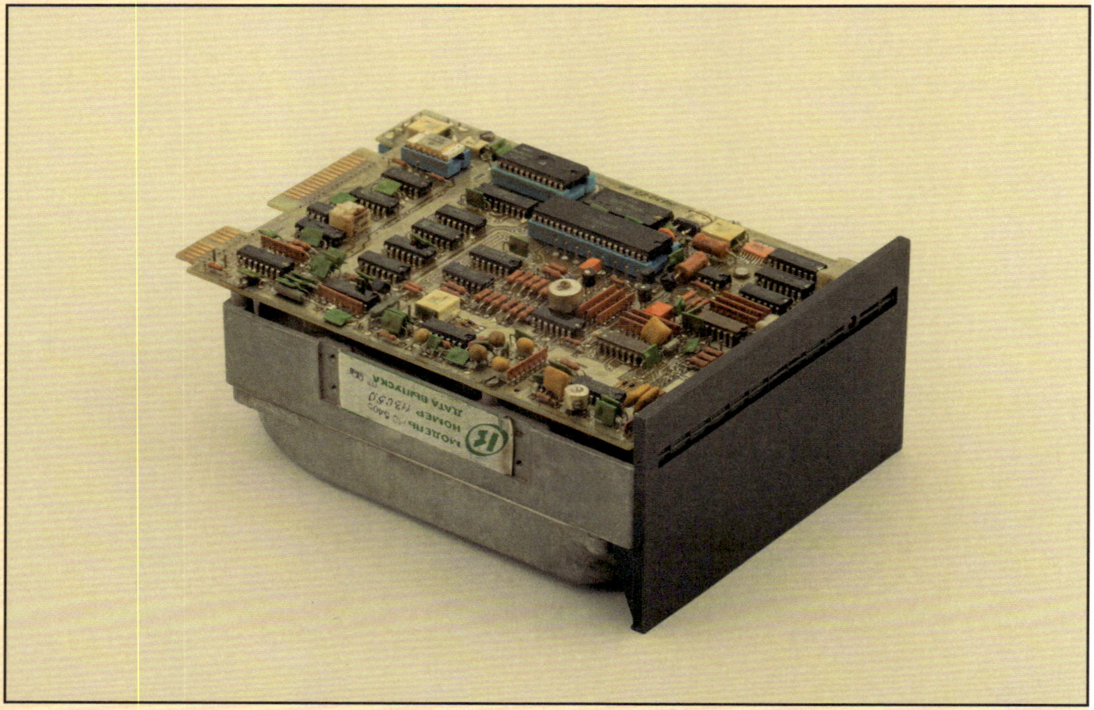

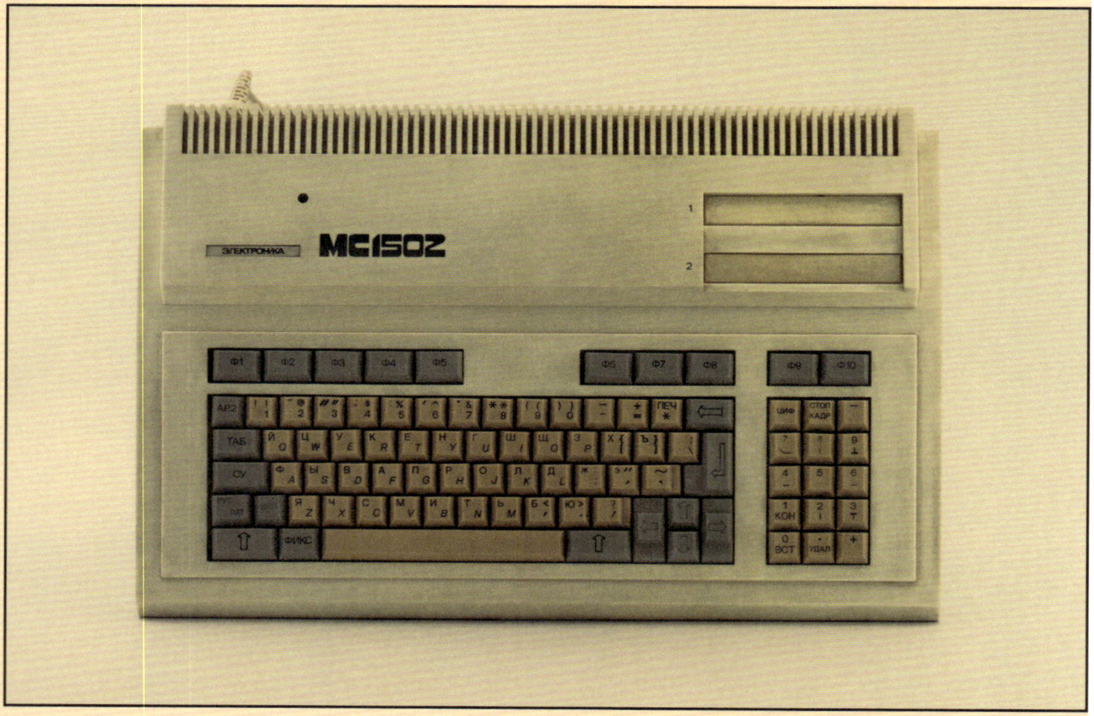

ЭЛЕКТРОНИКА
MC1502
1
2

Soviet MFM MS 5405 hard drive, 20 MB. Manufactured in 1990

creating the market for cheap IBM knock-offs in post-Soviet countries. With the real thing suddenly accessible to buyers in the former Soviet Union, the Poisk PC and its rivals soon vanished. Thousands of brand-new and used computers and computer accessories flooded the Ukrainian market. Ill-equipped to deal with the new reality, most Soviet enterprises ran out of steam. Not many home computers of that period managed to survive. During the economic winter of the mid-1990s, most PC owners took their computers apart and sold the precious metals inside for cash.

By the early 1990s, Electronmash had manufactured two new models of the Poisk. The Poisk 3 was manufactured in limited batches and the lifespan of this generation was cut short by the collapse of the Soviet Union. The main upgrade in the Poisk 3 involved simplifying and cheapening the design by using a larger number of microchips instead of semiconductor elements. Its specifications included an Intel 8088 clone processor (8 MHz in turbo mode), 640 KB RAM, enhanced graphics adapter (EGA), IDE disk controller, and a hard drive. The Poisk 3 wasn't the most powerful computer, but it was still fairly modern. Today, the Poisk 3 is highly sought-after by collectors for its rarity.

Most software developers now living and working in post-Soviet states learned first-hand about personal computers back in the early 1990s, during their Computer Science classes at school. After trying their hand at coding in BASIC or Pascal, they moved onto video games and got hooked.

A programmer recalls the days of computing in school: "I remember one of my classmates bringing a floppy with *King's Bounty.* We got so hyped up that we even snuck into the computer class after school for an all-night gaming binge."

Turn-based games like *King's Bounty* were extremely popular in Ukraine and other post-Soviet countries. Kharkiv-based programmer Serhiy Prokofiev went as far as to develop *King's Bounty 2*, an unofficial sequel to the original strategy epic produced by New World Computing. This under-the-counter game with a Russian interface went viral. Easily, more than half of all IBM-compatible PCs in the former Bloc countries had it installed. These simple games marked the dawn of the Ukrainian game dev industry, an industry that has since become synonymous with Ukraine. In the early 2000s, Prokofiev would go on to produce two versions of *Heroes of Malgrimia*, a tribute to his original hit.

Elektronika MC 1502 — a rival of the Poisk PC

THE GMD-130 FLOPPY DISKS

GMD-130, 5¼-inch floppy disks produced at Electronmash plant in Kyiv (below). 5¼-inch floppies were manufactured by a number of Soviet factories, but the GMD-130 model produced at Elektronmash was probably the most popular. The photo shows a double-sided disk with double-density recording that could store up to 720 KB of data. Compare this to the earliest 5¼-inch floppies manufactured by Shugart Associates in 1976, which could hold only 110 KB, or to the most advanced ones holding up to 1.2 MB. 3½-inch floppy disks were hardly ever used, since Soviet factories did not produce disk drives that could read them.

label, they had to use a felt-tip pen, never a ballpoint pen or a pencil, as those could damage the disk. In the photo, the opening that allows the drive's heads to read and write data is not covered with a sliding metal or plastic shutter like those used in 3½-inch floppies. On the label, 2×80 stands for "double-sided, 80 tracks on each side." The storage capacity of a floppy disk depended on the way it was formatted. Computers and operating systems used different types of formatting, so for the ZX Spectrum PC with TR-DOS, the GMD-130 floppy could only hold up to 640 KB instead of the usual 720 KB.

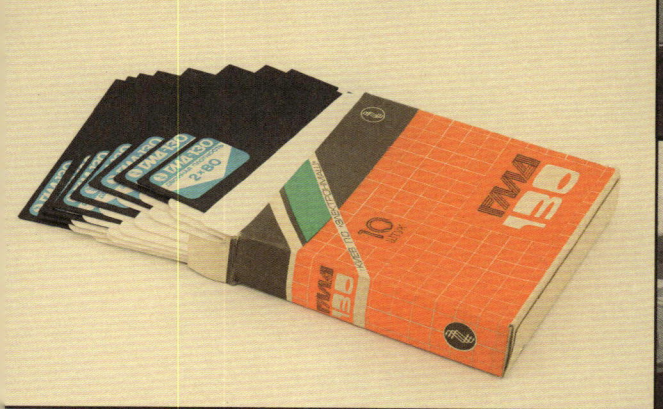

There was a small write-protect notch on the side of a floppy. When covered with the special sticker that was always included when buying a floppy, the user would be prevented from writing anything on the disk. A black plastic hub ring, installed on some floppies, helped reduce the wear rate. If someone wanted to write on a

MICROPROCESSOR KP580IK80A AND THE SM1800 PC

Ukraine was the Silicon Valley of the Soviet radio-electronic industry, so it was only natural that early clones of Intel microprocessors were launched into production at Crystal. This Kyiv-based microelectronics center was run by Alfred Kobylynskyi, a devoted, award-winning engineer and researcher.

Its Intel 8080 clone that came out in 1977 copied both its benefits and drawbacks. The Intel 8080 had a clock frequency limit of 2 H (2 MHz) and a 16-bit address bus, allowing easy access to 64 KB of memory. It used cutting-edge 6-micron NMOS technology and had exactly 4,758 transistors. One key thing about Intel 8080 is that it was the heart and the brain of Altair 8800, the first commercially successful — and affordable — personal computer. It was for this model that Bill Gates and Paul Allen developed their first commercial interpreter for BASIC. Intel 8080 was a trailblazer in the microcomputers era.

KP580IK80A, the Soviet clone with a mouthful name, was claimed to be an exact copy of the game-changing Intel 8080, but it wasn't actually. KP580IK80A was accurate, however, in replicating two major drawbacks of its Western counterpart: it required three power sources (-5, +5 and +12 V) and a separate chip clock generator. In 1976, Intel launched the Intel 8085, an updated processor that needed only one power source. Its Soviet clone appeared only a decade later and wasn't widely used.

The SM1800 model was the first Soviet PC to use the KP580IK80A microprocessor. The PC itself was also a clone of the Norwegian MYCRO-1, one of the first commercial single-board computers in the world. The SM1800 PC was a multipurpose machine used in automated process control systems, engineering, the energy sector, instrument manufacturing, transport, automated scientific research, and around the office.

The SM1800 was prototyped at the Controlling Electronic Machines Institute in Moscow and mass-produced at Elektronmash whose engineering department added the finishing touches before sending it to the assembly line in late 1980.

Customers could choose between several versions equipped with different types of software and computer accessories such as a Polish DZM-180 dot-matrix printer. For data input/output, the SM1800 used a video display terminal manufactured in the Ukrainian city of Vinnytsia.

The model was based on an operating system that copied CP/M, its American mass-market counterpart of the late 1970s created for Intel 8080 / Z80 based computers. From the get-go, the system offered a wide selection of pirated software — a huge perk at that time.

This photo (page 132, top right), taken at Kyiv Polytechnic Institute's museum, shows the SM1810 PC produced at Elektronmash in the late 1980s — early 1990s. Equipped with a 16-bit processor (a clone of Intel 8086), with 256 KB-1 MB random access memory, a 5¼-inch floppy disk drive, and a removable hard disk drive, it was much more advanced than its predecessor SM1800. This model was much more powerful, too, although it didn't really match world standards, since the original Intel 8086 had come out back in 1978, eight years earlier than its Soviet clone.

chapter 11

First-graders at Kyiv's School No. 79 during computer club. They are learning on Japanese Yamaha MSX computers, which were officially supplied to the USSR.

The USSR's Final Chapter

Elektronika BK 0010-01, a relatively inexpensive 16-bit computer, was a fixture in Soviet homes and classrooms. Over 160,000 BK series computers were manufactured. BKs, equipped with 3-MHz processors and 32 KB RAM, had their own operating systems that weren't compatible with similar foreign models.

The 1980s were decisive for the Soviet Union. The new decade dawned with UN sanctions in response to Soviet military action in Afghanistan. Ronald Reagan, a fierce opponent of communism, became president of the United States in 1981; a year later in the USSR, Leonid Brezhnev died after eighteen years as General Secretary of the Communist Party of the Soviet Union. His replacement, former KGB chairman Yuri Andropov, further worsened relations with the West with a series of foreign policy gaffes before dying of kidney failure in February of 1984. A short thirteen months later, his replacement was laid to rest, as well. A stagnating economy and unstable leadership in the Kremlin left many Soviet citizens feeling bewildered.

The appointment of a much younger and more progressive Party functionary as the general secretary of the Communist Party in 1985 brought fresh energy to the Union. His name was Mikhail Gorbachev. Seeking to modernize the economy, improve relations with the West, and establish a degree of freedom of speech, he launched a series of reforms that became known as *perestroika*. Unlike his predecessors, he thought it was essential to abandon rigid state planning and make a transition towards the market economy. He took decisive steps to modernize the country, namely in education and the IT industry.

Gorbachev's new policies yielded some good results. The Iron Curtain rolled up, giving the Soviet people the freedom of movement and a long sought-after chance to explore Western culture. Those with an entrepreneurial spirit got the green light to run their own businesses; many of them imported foreign goods. Western literature, journals, and academic books poured into the country. Perestroika held tremendous promise for the young generation.

In 1985, Gorbachev's government came together to discuss something that would have seemed unimaginable to the Soviet cybernetics pioneers like Lebedev and Glushkov: establishing an IT department in the USSR Academy of Sciences. The proposal also called for computer classes in public schools. The plan was quickly approved, and on September 1, 1985, Introduction to Information Technology and Computing became part of the national curriculum. Across all the Soviet republics,, school children began studying how to use computers. That's when things got complicated.

First, there were the shortages — of computers and of teachers. After decades of poo-pooing cybernetics as "bourgeois pseudoscience," there were hardly enough qualified instructors to immediately begin teaching millions of school children. The USSR was paying the price for being slow out of the gate on cybernetics. Engineers and programmers working at research centers stepped up and taught these classes, but although they all agreed that computing was valuable, the details were up for debate.

It took close to two years for educators to decide which programming language the students would use. Initially, they favored a Russian algorithmic language, although it was rather clunky. Then they alternated between Rapira and ALGOL for another two years. Later on, nearly every school switched to BASIC, which was ideal for students as a simpler language. Finally, Pascal was added to the curriculum.

At first, students learned to code from diagrams and books. Only after several years of this could leadership be persuaded that actual computers were a necessary part of the curriculum. Accordingly, Soviet factories finally began mass-producing the first personal computers. With access to foreign computer research, unified government support for IT, and computer training in schools, the decade showed all the signs of being a breakthrough for IT development.

Then, Chernobyl happened.

April 26, 1986, 1:23 a.m. It was the moment of the first explosion at the Chernobyl Nuclear Power Plant — and the beginning of the end of the Soviet Union. While the Kremlin downplayed the explosion to its citizens, poisonous fumes spread across Ukraine and into Europe, making Chernobyl a global threat.

This was the most disastrous nuclear accident in history, and the toll it took was not only on public health. The extent of the damage and the government's role in covering it up fomented distrust in the Communist Party. The USSR's reputation abroad suffered a heavy blow, as well. And across the USSR, the republics felt economic shockwaves that somehow never seemed to reach Moscow.

↑ →
Students at Kyiv's School No. 132 during a computer class at the Institute of Cybernetics' Computing Center

→ Students at Kyiv's School No. 132 during a computer class at the Institute of Cybernetics' Computing Center

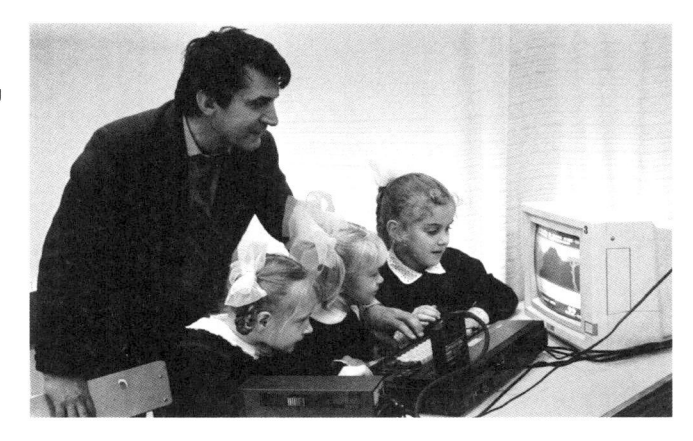

→ School students at a computer club in Kyiv. A 5¼-inch floppy disk produced by Elektronmash in Kyiv is in the foreground.

A meeting of the Young Naturalists Club. Sixth-graders from Sakhnivka middle school, Cherkasy region, Ukraine, are working on the Agat computer, compatible with Apple IIe.

The Agat-9 PC, targeted at the education sector. Compatible with Apple IIe, it had a hefty price tag, but in the early 1990s, it was available in electronics stores. It was equipped with 1 MHz MOS 6502 clone processor, 128 KB RAM expandable to 640 KB, and OS Apple DOS 3.3. From 1984 to 1993, nearly 150 thousand Agat-series PCs were manufactured.

The BK-0010 model with the Elektronika 302 VTC-201 monitor manufactured on the basis of a portable color TV; a power supply unit; a software cassette, a standard accessory at that time; and a Testing and Diagnostics Monitoring System unit.

The fallout of Party loyalty was more than the authorities could handle. Newspapers exposing the government's wrongdoings only facilitated a mass exodus of its members. From the republics, the trickle of sovereignty declarations that started in 1990 turned into a flood by the end of the year. Ukraine declared independence on August 24, 1991.

The republics refused to pay their contributions to the Union budget, which Moscow had always collected and then redistributed later across the Union. Ukraine, along with the other republics, found itself in dire straits. With the ruble nearly worthless, Ukraine began circulating coupons as an alternative currency. Soon, the republic's entire economy lay in shambles. Enterprises were shutting down right and left. Budget cuts were so severe that public-sector employees didn't get paid for months at a time. The country was plagued by mass unemployment.

In such a climate, government financing of knowledge-intensive industries was out of the question. The progress of cybernetics in these tumultuous years can be credited entirely to a handful of determined engineers who chugged away at the chipboard with little recognition and negligible reward.

For the former Soviet republics, the dissolution of the Soviet Union was not a simple victory. Ukraine — a newly independent country — had to shoulder the burden of Soviet mismanagement on its own, the aftermath of the Chernobyl disaster becoming a straitjacket for its budding economy. And what of the future of cybernetics? How the field would continue — indeed, whether it would continue at all — was far from certain.

chapter 12

↑
IBM-compatible computers at the Kyiv-Mohyla Academy in the early 1990s

1990s: After the Doldrums

Since 1951, when Kyiv's Laboratory #1 produced the first Soviet computer, Ukraine had been on the forefront of the technological revolution in the Soviet Union. As the final decade of the twentieth century arrived, the nation moved to the forefront of a different kind of revolution. On July 16, 1990, a little over a year before the Soviet Union's collapse, Ukraine adopted the Declaration of State Sovereignty of Ukraine. The Declaration boldly gave Ukrainian law precedence over Soviet law.

It wasn't yet independence, but it was a step out the door.

Four months later, students took to the streets in what has since been called Ukraine's "First Maidan," the Revolution on Granite. Protestors camped out in tents on Kyiv's October Revolution Square (today's Independence Square), calling for the nationalization of the Communist Party's property, Supreme Council re-elections, and a multi-party system. The occupation of Kyiv's central square has since become a recognizable image of revolution in Ukraine, constant in both the Orange Revolution in 2004 and the Revolution of Dignity in 2013-2014.

Hunger strikes among the protestors quickly created fractures in Ukraine's Supreme Council. Laughter from some deputies on hearing the news prompted the departure of others, like Ukrainian author and deputy Oles Honchar, who rebuked his colleagues, "Those are our children!" His departure spurred a mass exodus of ordinary Ukrainians from the Communist Party and finally, concessions from the Soviet government. The Revolution on Granite ended on October 17, but the quiet of Kyiv's streets didn't last for long.

On August 24, 1991, the Supreme Council of the Ukrainian Soviet Socialist Republic rechristened itself the Verkhovna Rada of Ukraine. The body declared independence from the Soviet Union and the creation of an independent Ukrainian state. A country-wide referendum held on December 1, 1991 of that year confirmed this decision.

90% of participants supported the Ukrainian independence. On December 8, the USSR ceased to exist officially.

Politically, the country was experiencing growth. Economically, however, the decade was a period of crippling

Tents in Independence Square subsequently became the most recognizable symbol of revolution in Ukraine. They appeared here again in 2004 and 2013.
↓

↑
In the early 1990s, Ukrainians were allowed to engage in free market enterprise. Little grocery stores, even mobile ones, opened right and left. These young people are selling a small selection of staples out of the back of an old ambulance van.

↑
The Central Republican Stadium had endless rows of stalls. Newly-minted entrepreneurs — former school teachers, university professors, doctors, and engineers — started buying foodstuffs, clothing, and other goods from former Bloc countries, Turkey, and China, and selling them in Ukraine.

economic stagnation. The decadeslong ties between the former Soviet states were severed, forcing hundreds of enterprises to shut down entirely or to dramatically curtail their operations. Inflation soared to 1000% between 1992 and 1994. State employees went without paychecks for months, and then months turned into years.

For many Ukrainians, it felt like their world had been turned upside down. School teachers, doctors, university professors, and engineers had the toughest lot. In the Soviet Union, these positions were all public, without exception. Without private infrastructure to transition to, many of these professionals turned to the only paid work available: private entrepreneurship, typically in the form of curbside stands of imported cigarettes and minor household goods.

Shuttle traders, who made trips to Poland, Turkey, and Bulgaria for their products, were in high demand. Spontaneous markets flooded the country's cities. Kyiv's 70,000-seat Central Republican Stadium (today's Olimpiyskiy Stadium) hosted the largest of them. There, the sight of a former biology professor selling stockings from Poland alongside a math professor hawking blouses from Turkey was nothing to raise eyebrows. For former Soviet citizens accustomed to a planned economy, it was a bewildering time.

At the same time, there were those who were in a position to pursue greater opportunities in the new market economy. In those years, fortunes were made swiftly. New firms, newspapers, and television companies sprung up overnight. The period ushered in something completely new and for some, new wasn't half bad.

Now, back to cybernetics. By the early 1990s, the once-grand Soviet cybernetics industry had been leveled. Research had come to a screeching halt; new computers were no longer being designed. Up until 1993, Ukrainian-designed PCs managed to keep a foothold in the market. Soon enough, however, IBM PC clones and Macintosh computers, which were already in high demand among a limited number of well-to-do designers and experts in the printing industry, asserted their dominance on the Ukrainian market.

Customer preferences shifted from Ukrainian analogues of the ZX Spectrum to Dendy, a popular pirated clone of the Nintendo Entertainment System. SEGA, Japan's more expensive gaming system, also had a presence in Ukraine. Sony PlayStation joined them slightly later. The final nail in the coffin of the Ukrainian computer industry was the lifting of restrictions on Western tech imports to former Soviet states. By 1994, the original designs by Puyda, Volkov, Popov, and other Soviet-era engineers had ceased to be produced.

But there was an upside. Ukrainian factories suddenly had access to inexpensive PC parts imported from Asia. At that time, computer manufacturing was a Wild West of firms, amateur entrepreneurs, and reincorporated Soviet factories assembling computers for resale at stores, radio markets, or through newspaper ads. Some of these businesses, like microprocessor factory Kvazar-Micro, resembled Dell in its early years of rapid expansion.

In these unstable times, Ukraine's IT industry boiled down to four activities: distributing foreign products, complete knock-down assembly, developing specialized software (mostly accounting programs for domestic use), and selling pirated CDs. Projects of another nature were the exception, not the rule.

This is a typical IBM PC clone assembled in the former Soviet Union. The computer has an Intel 80386 33 MHz processor, which was popular in the early 1990s. Disk drives for 3½-inch and 5¼-inch floppies were quite common throughout the 1990s.

Copy culture, or piracy, bloomed in the 1990s due to flawed copyright laws, pervasive corruption, poverty, and total ignorance about the way intellectual property should be handled. The concept of piracy as a criminal act did not yet exist. Because of the prevalence of cheaply produced bootlegged versions of hit games from Asia and the U.S., there was hardly any way someone could make money producing high-quality software for the Ukrainian market in the 1990s.

If the Ukrainian market looked like a pirate stronghold, it was because the Western one was simply out of reach. In the newly minted republics, the Internet was still in its infancy. Most people barely spoke English; Western cultural norms were not understood well. In business dealings with Western companies during this transition period, misunderstandings were the norm rather than the exception.

Some companies made it really big. GSC Game World is a rock star of the Ukrainian game dev that shot to international stardom back in 2006 with its flagship *S.T.A.L.K.E.R.*, set in the Chornobyl Exclusion Zone. Even its earlier products — *Cossacks*: *European Wars*, *Codename*: *Outbreak*, and *American Conquest* — were popular and profitable. The company became a trailblazer for the emerging game development industry.

An enterprise in Kyiv that used to copy CDs in bulk in the mid-1990s. Companies like this one often produced both branded and pirated discs. They also churned out "shadow license" discs — illegal copies that looked very similar to their branded counterparts. When audited, pirates always had papers at hand that were good enough to keep the government off their backs.

← The photo shows Classic 2, a modified version of the original Dendy.

↓ This photo shows Junior, another Dendy iteration. Its case was almost an exact copy of Nintendo Famicom.

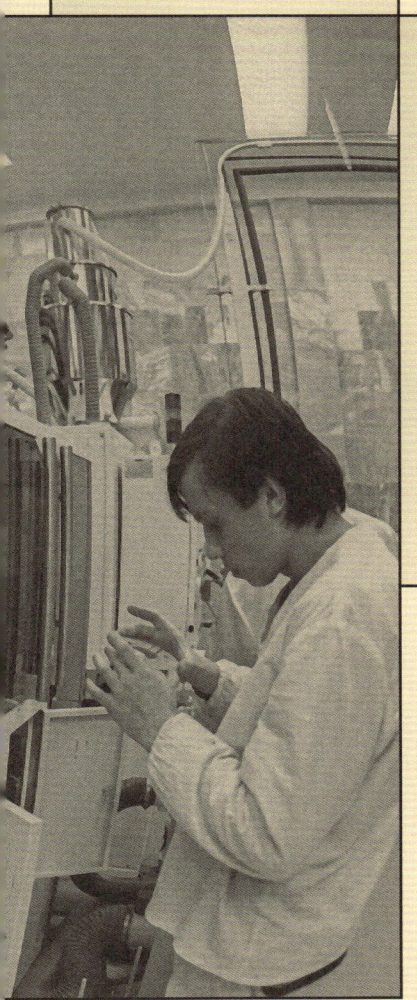

Dendy, an illegal but fairly good and super popular NES clone. This game console was manufactured in Taiwan and widely sold in post-Soviet countries. There were brand stores and even a brand magazine. Naturally, most of the customers were sure they'd bought legal consoles.

GSC GAME WORLD AND S.T.A.L.K.E.R.

GSC Game World was set up in 1995 by a sixteen year-old boy who conceived his future startup while still in high school. Initially, S.T.A.L.K.E.R. — the company's flagship first-person shooter and its greatest achievement — was expected to appeal to Western gamers, but later on, the team decided to showcase something unique. They picked the Chornobyl Exclusion Zone as a setting, carefully recreating the abandoned town of Prypyat and the surrounding landscapes.

The game blew the global game market away, selling 5 million copies and generating a profit of $100 million. For the Ukrainian game dev, it was an unheard-of success. The company's CEO, Sergiy Grygorovych, received the EY Entrepreneur of the Year award in 2009. GSC Game World has released several new installments in the series and released a much awaited sequel in 2021.

← The first-generation Sony PlayStation that hit the market in 1995. It was no comparison with Dendy or even SEGA.

→ SEGA MegaDrive 2 or SEGA Genesis, and a cartridge with *The Lion King*. Both were big hits in the mid-1990s.

Ihor Karev, Action Forms co-founder and creator of first-person shooter game *Chasm: The Rift* (1997), says: "Those were golden times for Ukrainian game development. The industry was only just emerging. A lot of young talented guys were busy developing their dream games, treating this like their hobby rather than a business. The talent pool filled quickly, startups sprang up across the country.

"Ukraine made a name for itself in the global game dev market as a nation of creative professionals. Western companies eager to tap into its potential set up development offices here. I got caught up in this hype, too. The fact that my English wasn't great or that I had no idea how the market functioned didn't hold me back. I just started to look for an investor for the game I'd been developing.

My business partner and I — he was twenty at that time; I was eighteen — just dropped by the offices of local IT companies until we finally reached Global Logic, probably the largest Ukrainian internet provider at the time. They told us, 'We can't give you any money, guys, but we can offer you computers and desks so you can come here and work on your game. If something comes out of it, we'll discuss how we could go about it.' That's what we did."

2

And what of Ukrainian innovation in the 21st century?

That story is far from finished. These eight global tech brands, headed out of Ukraine, are just the tip of the iceberg.

MacPaw: Empowering Humans with Technology

← Oleksandr Kosovan, Founder and CEO of MacPaw

MacPaw, a maker of award-winning utility software for Mac computers, is one of a handful of Ukrainian tech startups that was born from ideals. With more than 30 million users, the company's strong focus on design and functionality aims to create a world where technology empowers human life. MacPaw's people-oriented products are designed for impact and have been recognized with a Red Dot Award, a Golden Kitty, and other awards, with its top app, CleanMyMac. But the company's most impressive feat lies in its origin story.

↑
CleanMyMac X
app by MacPaw

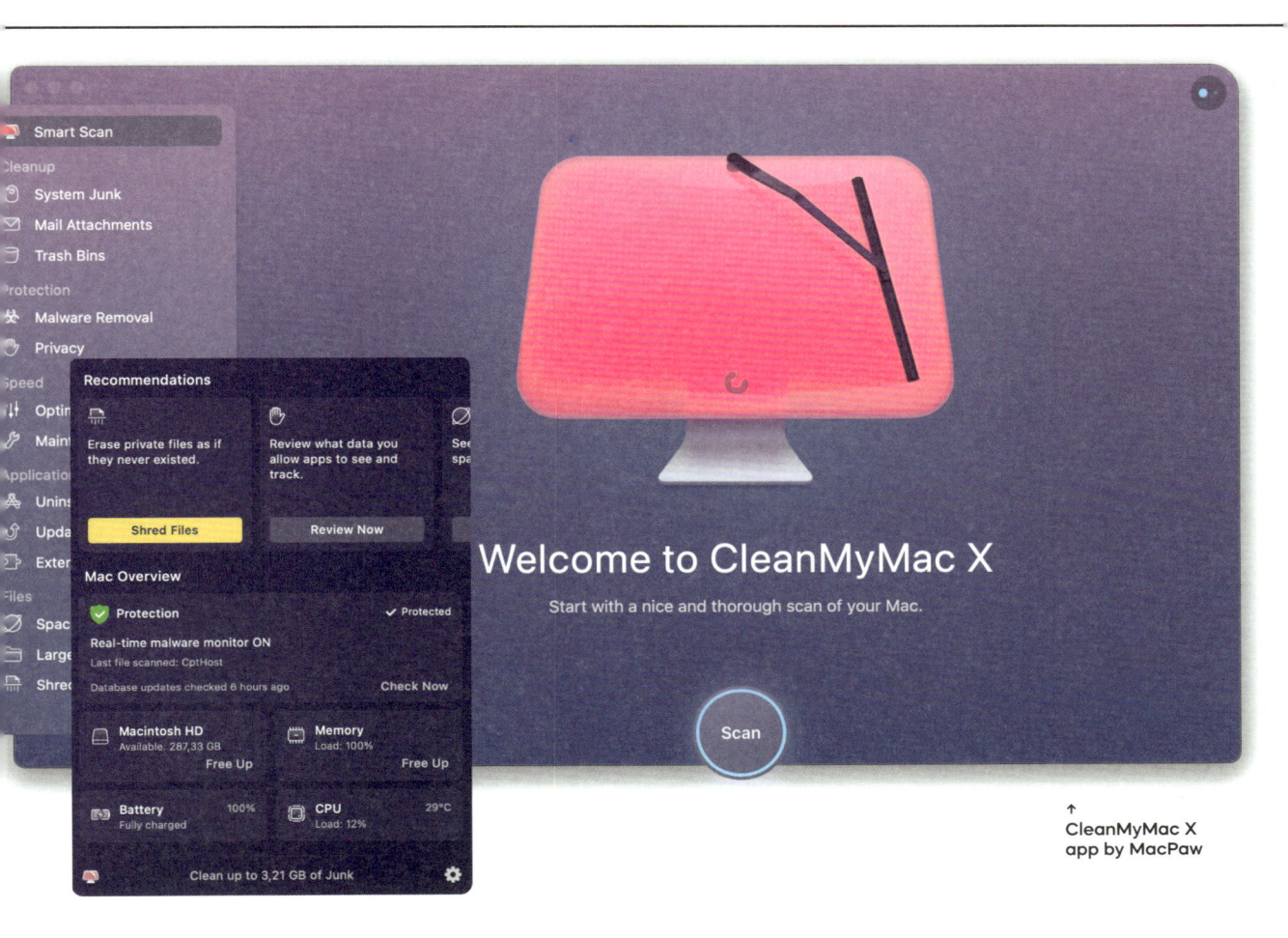

MacPaw was once just a small team of a few people. Today, its software is run on one in every five Macs in the world. How did they do it? MacPaw's story starts in Kyiv with Oleksandr Kosovan, a self-made innovator who achieved success in one of the most competitive Western markets by putting two things first: love of Apple computers and the belief that technology should simplify everyday life. Let's back up a little bit first, though.

Oleksandr comes from a family of engineers. At school, he got a summer job helping his father at a construction site in France. While still a student at Kyiv Polytechnic Institute, Oleksandr worked as a Windows system administrator but then switched to Unix systems. He spent his first paycheck on an Apple keyboard. "My love for Apple products began when I was a freshman in college. I stopped by a store and saw the iMac (flat panel) — some people call it the iLamp. The computer came with a see-through, wireless keyboard with white keys," he remembers.

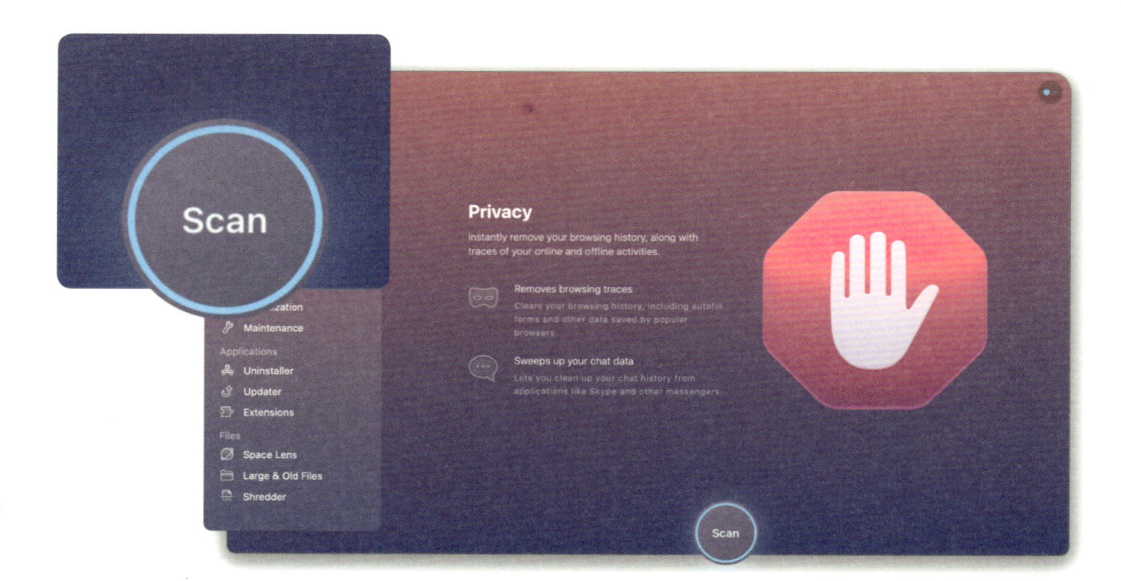

→ Privacy check-up and "Scan" button in CleanMyMac X

Then a student, Oleksandr thought long and hard about the purchase. "I was short on cash, as it was worth my whole monthly paycheck, but I was sure I had to buy it! Eventually, I bought the iMac keyboard. This was my first Apple product and maybe the best purchase of my life." It was the start of an obsession. "I took an interest in macOS, started following Apple's new releases, watched Steve Jobs's keynotes, and saved up for my first Mac," he says.

All of Oleksandr's hard work resulted in CleanMyMac, the flagship product of MacPaw, his newly formed software startup. Many, if not most, companies reach success only after a string of failed or less-than-stellar projects. With MacPaw, its very first product is what earned it a spot at the top. CleanMyMac proved to be exactly what its audience wanted. Its high quality keeps CleanMyMac in high demand, and the product is still at the core of the business.

Among apps for macOS, CleanMyMac, one of MacPaw's many products, is in a class of its own. The product combines the capabilities of many computer optimization tools to keep Macs clean and healthy. CleanMyMac enables users to find and delete tons of unnecessary flies buried in the deepest corners of their file systems. At the same time, it continuously monitors computer performance, detecting and stopping harmful software.
All of that just by clicking a single button!

152

> *"Attention to detail and fine-tuning the little things — what most teams think is unimportant — is what differentiates top-quality products from mediocre ones. The central "Scan" button in CleanMyMac is one of those little things. The button is tremendously powerful and so many of the program's functions are tied into it. Each thing that it does was seamlessly integrated by the developers behind the scenes. At the same time, the button protrudes beyond the edges of the app, which makes it hard to copy. That button became our distinguishing feature, and trying to replicate it would have involved a great deal of effort. Basically, it became our seal of quality that distinguished us from the numerous imitators that had popped up after we hit the market,"* says Oleksandr Kosovan.

Oleksandr's success strategy was to bootstrap MacPaw in the company's early days. At first, he did almost all of the work himself, investing his time and energy. "Coding, product design, website development, server support — that's what I knew and was good at," Oleksandr recalls. "I wrote all of the code for the first version of CleanMyMac. It took me almost a full year. I mostly worked at night or on my days off. Well, actually, I didn't really have any days off. I did have a goal, though — making a good product and going to market with it. I coded for the first five years of the company's existence."

However, his focus eventually shifted to managing the company, growing the team, and creating new products. Now, technical work is off Oleksandr's plate completely. "I really miss coding, as you can immediately feel the results of your intellectual work. You're thinking about something, trying to solve a problem, and then — bam! — you come up with a solution, and everything works! It's an indescribable feeling of euphoria, and I don't get enough of it as a manager."

Of course, people who excel at absolutely everything simply don't exist. Oleksandr enlisted outside help for product and UX design when developing the second version of CleanMyMac. After this, the company received its signature logo in the form of C-shaped cat paw. At the time, cat-themed naming was something of a trend for the macOS market. MacPaw joined the ranks of start-ups with names like Tiger and Leopard and took the trend a step further by making its office home to two actual cats, Fixel and Hoover.

While developing CleanMyMac, Oleksandr decided to start making a second product — MacHider, a file privacy app. Some might wonder why a budding entrepreneur with a new product would disperse his efforts like that. If you talk to Oleksandr, he will tell you that the move was well calculated.

> *"Actually, it was part of a strategy to increase the average sale amount that I'd seen somewhere. For instance, if you're selling a product for $30, you can offer another product, which customers might not actually need at that moment, at a discount. Many customers go for it. I thought it wouldn't be a bad idea to use that strategy and offer users something extra."* That second product was MacHider, which Oleksandr worked on together with a former classmate, Vira Tkachenko, now MacPaw's Chief Technology and Innovation Officer. *"From the early days, Vira has become a core partner in many areas. She is responsible for our company's technology and innovation culture,"* says Oleksandr.

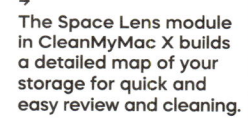

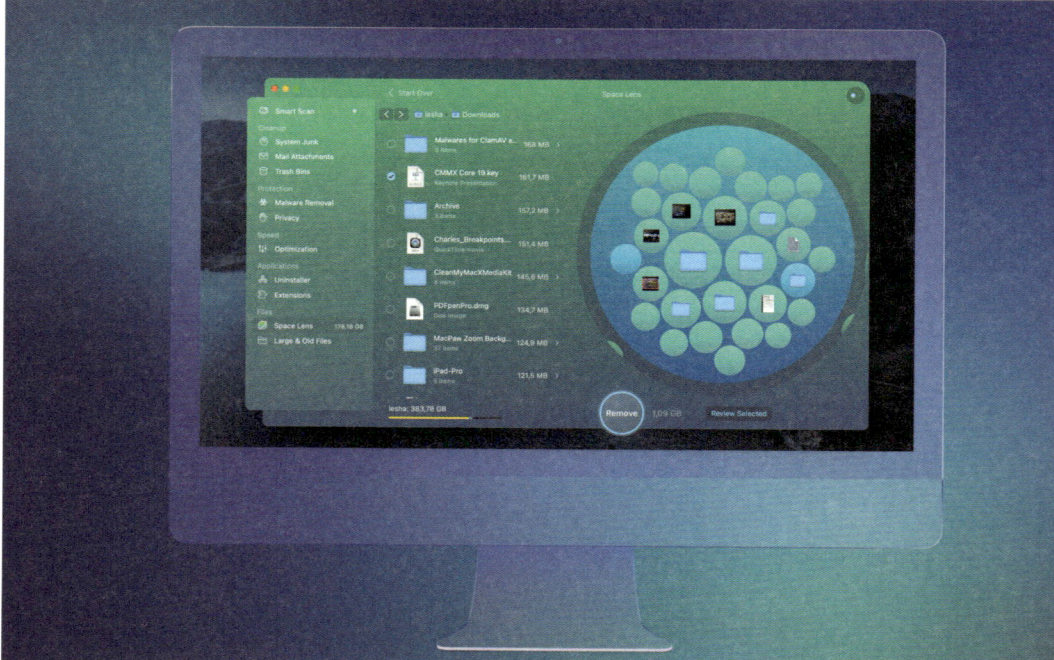

→ The Space Lens module in CleanMyMac X builds a detailed map of your storage for quick and easy review and cleaning.

The first versions of CleanMyMac and MacHider hit the international market in January 2009. It started selling well right away. Oleksandr still remembers "the inexpressible thrill" of those first months. "It is a fantastic feeling when people are actually ready to pay for what you have created and have been working on for a long time. I get the same euphoria every time we release a new product or service for the customers. It was never about the money; it was about the feeling of grateful customers that love your products and are ready to express that feeling with a dollar," says Oleksandr.

Now, Oleksandr can take credit for more than making a few sales. He created a high-demand product and made it the foundation of a strong company. "I find it really hard to imagine how I'd replicate what I did at that time because my priorities have changed completely. I have a family, personal hobbies and activities. Besides work, my time is now divided between other areas of interest. In the early days, I could easily devote all of my free time to some hobby or a project that might not even deliver results. Now, I'm afraid I don't have that luxury," Oleksandr says.

The founder's view on entrepreneurship is that each company's development model "is tightly linked to where an entrepreneur is in their life." If they are able to invest time generously, then bootstrapping a project can be feasible. As for himself, he says building a business without resources nowadays is "hard to imagine." The reason? "Companies are fighting fiercely for talent. Generally, you either have an amazing idea that enables you to attract talent, or significant funding."

"Seeing things through is the hardest part. There are so many times when you want to stop, like when you've done 90% of the work — and that took a good chunk of time — and you suddenly want to quit and get a stable job, stop messing around. Unfortunately, a lot of projects die at this stage. Instead, try to take the product to a logical conclusion. That way, you'll see some results and not just throw all of that work away."

— Oleksandr Kosovan

As the company grew, it has continued to roll out products for Mac and iOS, following the success of CleanMyMac and MacHider. Among them are Gemini, a duplicate file finder, and Setapp, a subscription-based app platform to help users stay focused and organized.

Setapp stands out as an ambitious product — not just for MacPaw, but for the whole market. Launched as an alternative to Apple's App Store, Setapp introduced a more flexible way to distribute and use apps. For just $9.99 a month, users get access to 240+ apps for macOS and iOS, including MacPaw's own tools. All the apps provide access to premium features, and updates are free. A recommendation system helps users discover more useful apps and streamline their workflow. With new apps being added regularly, Setapp has become a one-stop platform for solving every task on Mac and iPhone.

In 2017, MacPaw added The Unarchiver, purchased from Finnish developer Dag Ågren, to its product line. "We had been watching The Unarchiver for a while. It was on the App Store's Top 10 Most Downloaded Apps for several years in a row," Oleksandr says about the decision to acquire the app. The Unarchiver is ubiquitous in the macOS community. "There's no going without it. We've been trying to create something similar with all of our products. We wanted The Unarchiver to be associated with our brand, so users who know The Unarchiver will use our other products, too."

Throughout its history, MacPaw has focused on growth, performance, and longevity. In a niche where competitors are constantly coming on the market, easing up on quality and innovation isn't an option. At the same time, MacPaw's leadership never strays from the company's mission: helping machines help people. As of 2023, the MacPaw family of apps numbers more than ten, including ClearVPN, SpyBuster, and other apps, each of them designed to simplify life with technology.

MacPaw continues creating new products and enhancing its active software with features that ensure a safe online experience. Just weeks after the start of a full-scale Russian invasion of Ukraine in 2022, MacPaw engineers created SpyBuster, a free, on-device anti-spyware app. It enables Ukrainians to protect their online data from leaking to unwanted servers and law enforcement agencies in Russia and Belarus.

In 2023, MacPaw launched Moonlock, a new cybersecurity division focused exclusively on the security of Mac users. The new division will work on creating easy-to-use cybersecurity tools for all Mac users. The first solution on the list is Moonlock Engine technology, which now empowers the Malware Removal module of MacPaw's most popular product — CleanMyMac X.

That's not all MacPaw's leadership team has its eyes on. "We periodically think about purchasing other products, too. But buying a product is only half of the process, since it takes a lot of resources and effort to support it. When you purchase a product you take it upon yourself to keep it up to date, work with its current audience, develop new features, and continue to improve it. So, that's a lot of responsibility, and we're not always willing to take it on, even if we get a good offer."

Over the past few years, there's been some grumbling from developers about the App Store. What does Oleksandr Kosovan think about all of that?

"First of all, the App Store guidelines and process is ruther unfriendly for many developers. They invest a lot of resources, including their time, to create a product only to find out that once they finish, their app might not even be accepted. They might be asked to change everything or even be forced to abandon it altogether. There are some advantages. Apple is keeping the platform "cleaner" this way, but developing apps for the App Store is a major headache because of these uncertainties."

— Oleksandr Kosovan

→
Setapp, a subscription-based productivity platform for macOS and iOS by MacPaw

In an industry where companies like to tout hip office spaces, there's only one word to describe MacPaw's striking premises: cool. Oleksandr says the office has certainly become part of the company's appeal to new talent, but it's a by-product of succeeding in their original aim for the space. "Our goal was not to lure people with a picture-perfect design, but to create an inventive, comfortable vibe that would encourage our people to work to their fullest potential. If you spend one-third of your life in the office, you deserve to do so in comfort."

At first, the space was nothing but concrete. Mykola Gulyk at Baraban Architecture Crew developed the design, including many ideas contributed by the team. "We paid special attention to the details. Meeting pods, coffee machines, other little everyday things. Someone came up with a bigger idea, to set up a museum of computer equipment — which we did, too," says Oleksandr. "Mykola's got a great eye for selecting materials, matching colors, and choosing decorations. We started out thinking about the kind of workspace that would be the best fit for MacPaw. Then, we added our people's ideas to this skeleton, mapped out the interior design, selected good furniture — and got a result we can be proud of."

MacPaw's office is a vivid example of the company's values in action: make impact, create experience, and stay human. "Personally, I'm very concerned about quality and I'm always trying to pick the best things, the best practices, the best talent," says Oleksandr. "That's a very basic example of our approach to other, bigger things."

← MacPaw team in a meeting room. MacPaw office, Kyiv.

Photographer: Katia Akvarelna for Ukraïner

Then, of course, there's the technology MacPaw uses. Every MacPaw team member is set up with an Apple computer, a detail Oleksandr insists on because of the brand's reputation for reliability. "It means that our people can focus on their projects instead of fixing their laptops. We're always on the lookout for things that can save time and boost productivity. And it shows in every aspect of our work."

← Scientists with satellite antenna presented by MacPaw at the Ukrainian Vernadsky Research Base, Antarctica, 2021

Apart from its products and work environment, MacPaw prides itself on its commitment to social responsibility. In 2018, the company launched its first social initiatives, contributing funding and volunteer hours in key areas under the umbrella of MacPawCares. Whether it's tutoring students in programming, distributing hot lunches to the elderly, or sprucing up Kyiv's green spaces, MacPaw's talent is always eager to pay it forward in their communities. The Sortui ("Sort It") app is one project that brings together MacPaw values of tech and stewardship. Ukrainians can use the app to properly sort household trash and find their nearest recycling station, making eco-conscious living that much easier.

Some of the company's social initiatives go beyond local impact, matching its global presence. The MacPaw flag has been fluttering in Antarctica since May 22, 2019, next to the Ukrainian Vernadsky Research Base, where a team of researchers study critical environmental indicators year-round. MacPaw embarked on an ambitious project to donate and ship a satellite antenna to the team.

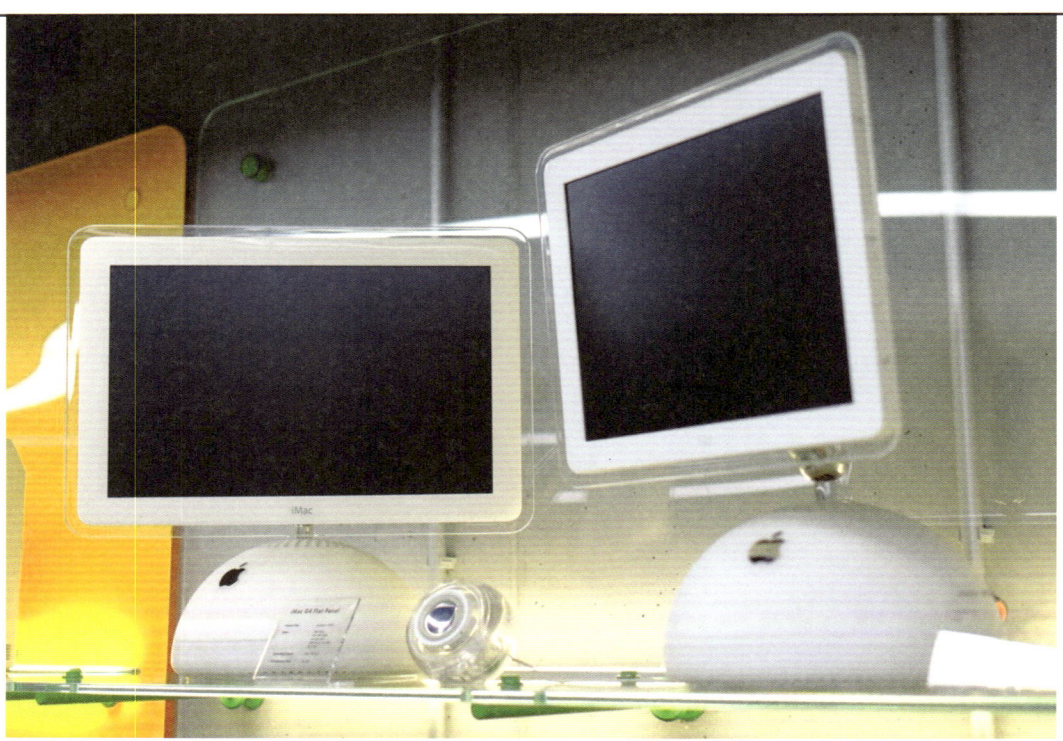

← iMac G4 Flat Panel (2002) and iMac G4 (2003) in MacPaw's collection of Apple hardware in the MacPaw office

← Welcome desk at MacPaw Space, a new creative platform in Kyiv, Ukraine

A more remote location here on Earth could not be imagined, but against numerous obstacles, the project was completed in 2020. Now, the research team can connect to the rest of the world eight times faster than the previous satellite antenna allowed. MacPaw has also provided the researchers with a number of rugged laptops to withstand the extreme Antarctic conditions, Wi-Fi components for the coputers, and additional research equipment. MacPaw team continues support the researchers even throughout the full-scale Russian invasion.

Back in Kyiv, the MacPaw team is working on a project closer to home, this time to cull inspiration from computing visionaries of the past and fuel innovation here and now. Part community center and part museum, MacPaw Space is a creative space for like-minded people to connect, share ideas, and drive change to the world. The sleek premises will welcome bright minds, changemakers, inventors and, of course, Apple geeks. MacPaw Space hosts meetups, lectures, workshops, and more. Complementing its main presentation stage at the Horizon amphitheater, lecture rooms and open plan meeting spaces, a permanent exhibition of vintage Apple hardware will give visitors a rare look into the evolution of personal computing from the 1980s through today.[1]

1/ MacPaw Space is independent and has not been authorized, sponsored, or otherwise endorsed by Apple Inc.

The museum will showcase 323 of Apple's most iconic products, including a Macintosh 128k signed by Steve Wozniak and special finds from a private collector.

"Some of them are just like pieces of art and we're very glad to have this opportunity to share these iconic items with other Apple fans," Oleksandr says of his company's collection.

So, what does the future look like for MacPaw? Oleksandr envisions a product company that continues to keep impact and empowerment front and center. "We want to empower people to do what they have a passion for. We want to make people feel able, reach a flow, be productive," the founder says. MacPaw's overall commitment to helping people befriend technology will continue to underlie its product development. With technology as a partner, each person's impact is nearly unlimited. Moving forward, the tools MacPaw builds will, just like its ambitious community projects, continue to evoke the special energy that fuels MacPaw's work: the feeling that the possibilities are endless.

Readdle: "Thank You for Trusting Us Over 200 Million Times!"

← Top row: Alex Tyagulsky, Dmitry Protserov.
Bottom row: Andrian Budantsov, Igor Zhadanov.

It's far from easy these days to make a mobile app that can really compete on the App Store and on Google Play. You need to have ample opportunity, as well as generous funding. Who would have predicted that a startup in far-flung Odesa, Ukraine would come to claim a hefty piece of the American market?

↓
Screenshots from Readdle apps PDF Expert and Documents

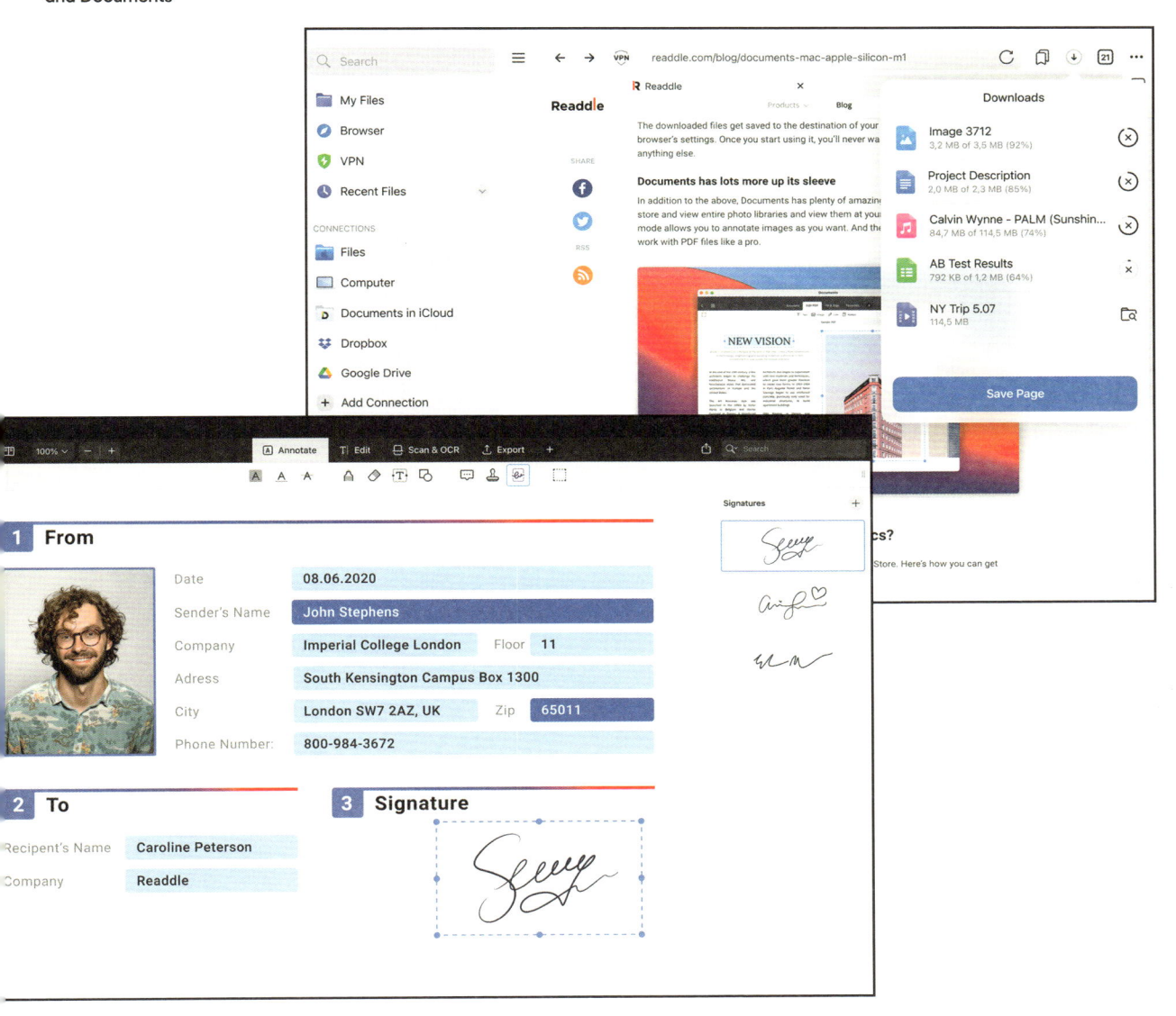

"Our parents have always supported us and tried to give us everything they can. In 1997, we took a major risk and spent a year's salary on a computer. Looking back, I can say that was the best investment our family ever made."

— Igor Zhadanov, CEO and Co-founder of Readdle

In 2007, four young programmers from Odesa, a charming, energetic city on the Black Sea coast, decided to start their own company. They planned to develop an app for the first iPhone, which had just come out. The App Store did not yet exist, so smartphone users had to content themselves with whatever pre-installed built-in apps came with their phones. However, the four friends from Odesa realized that would soon change. Soon enough, they had brought Readdle, a mobile app startup with one salaried employee, to the market.

> *"My partners and I were pretty sure we wanted to create our own product, so Readdle, in one form or another, was simply inevitable."*
> — Igor Zhadanov

So, what was their first app supposed to do? Turn any smartphone into an e-book reader. Rebranded as Documents, the app is still widely used today. It allows users to work with any type of document from their mobile device — and it's free.

Today, Readdle is a team of over 300 top-notch professionals who work in over 30 different countries. Their apps have been downloaded over 220 million times. Sixteen years since their first product launch, Readdle has managed not only to stay afloat, but to roll out even more top-quality apps that users love. Despite having grown tremendously over the years, Readdle has maintained its independence and avoided being absorbed by a Big Tech giant.

> *"The company's strength is in its independence and flexibility, which enables us to make the products we want to make."*
> — Dmitry Protserov

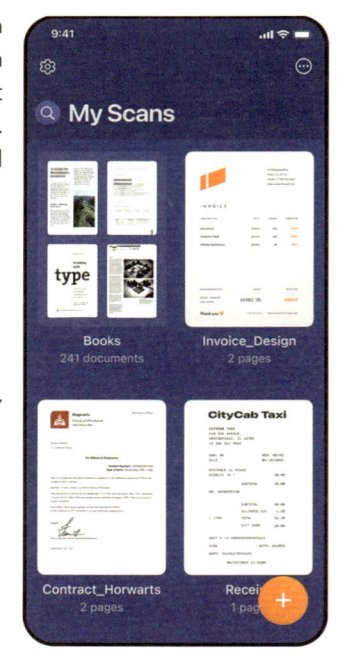

↑
Scanner Pro, on one of Readdle's apps

↓
Readdle's PDF Expert is packed with powerful features in a well-designed and intuitive interface

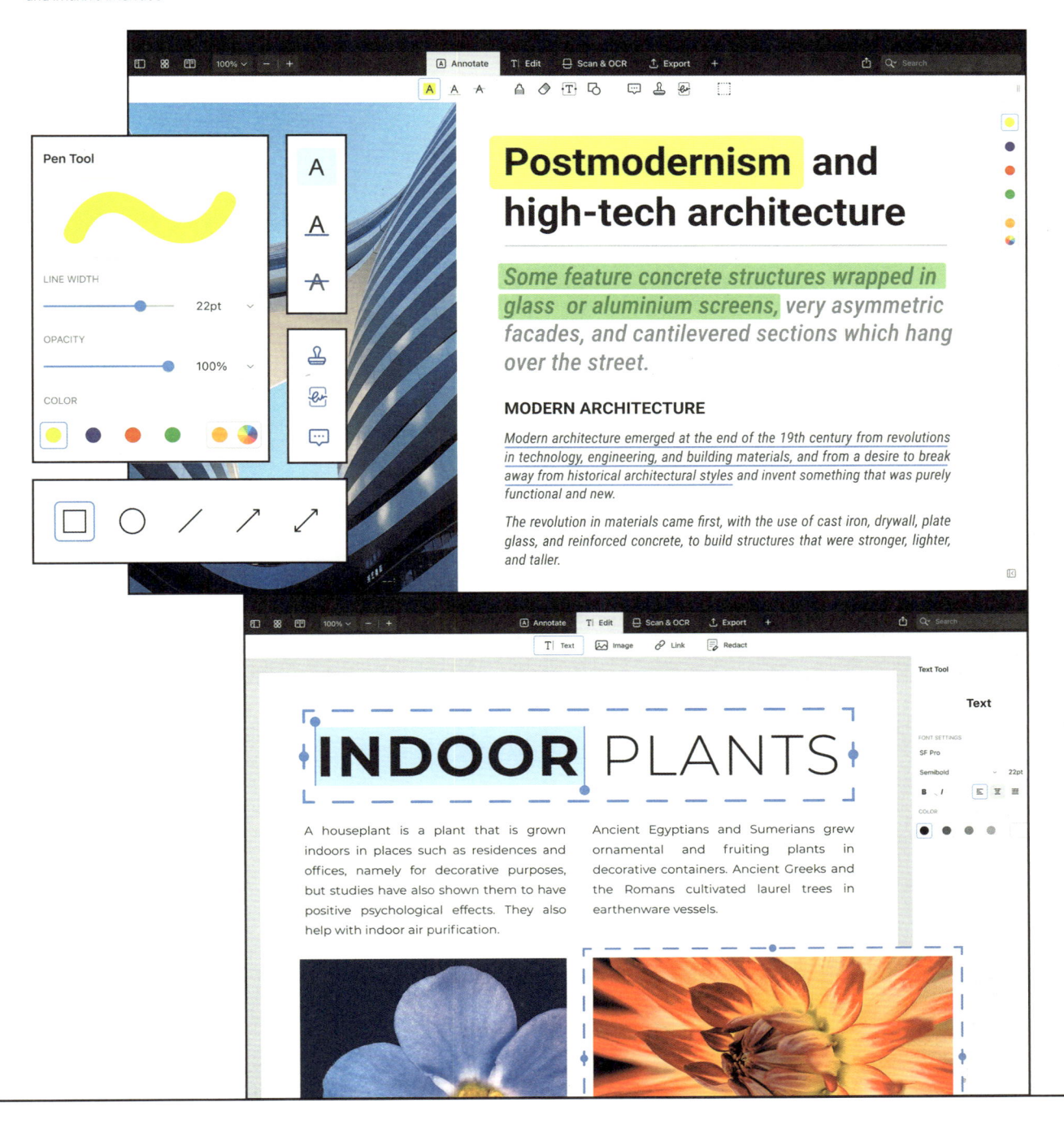

Skill and luck weren't the only two factors involved. Readdle's team has demonstrated remarkable persistence: out of the 40 products they've created, 32 have failed. But alongside the failures, their successes shine brightly.

> Documents, PDF Expert (2015 App of the Year Runner-up by Apple, 2023 Red Dot Award for design), Spark (Best of 2016 by Apple, 2023 winner of Golden Kitty by Product Hunt), Scanner Pro (Editors' Choice by Apple), and Calendars (2021 Red Dot Award: Product Design) by Readdle are among Readdle's leading apps. The purposes of all of these apps are obvious enough from their names, except, perhaps, for Spark.

Spark is the first "collaborative email" tool, designed to make people love email again. Designed as a universal communications tool, it brought many new features to the market. Now ubiquitous features like adding private comments in a group chat, email collaboration, and external linking to emails or chains all originated with Spark.

> The idea for Spark came from a question: Can email be significantly improved, or are new iterations inevitably just reinventing the wheel?

At Readdle, the answer was obvious. How could a platform that's stayed more or less the same for 25 years *not* need improvement?

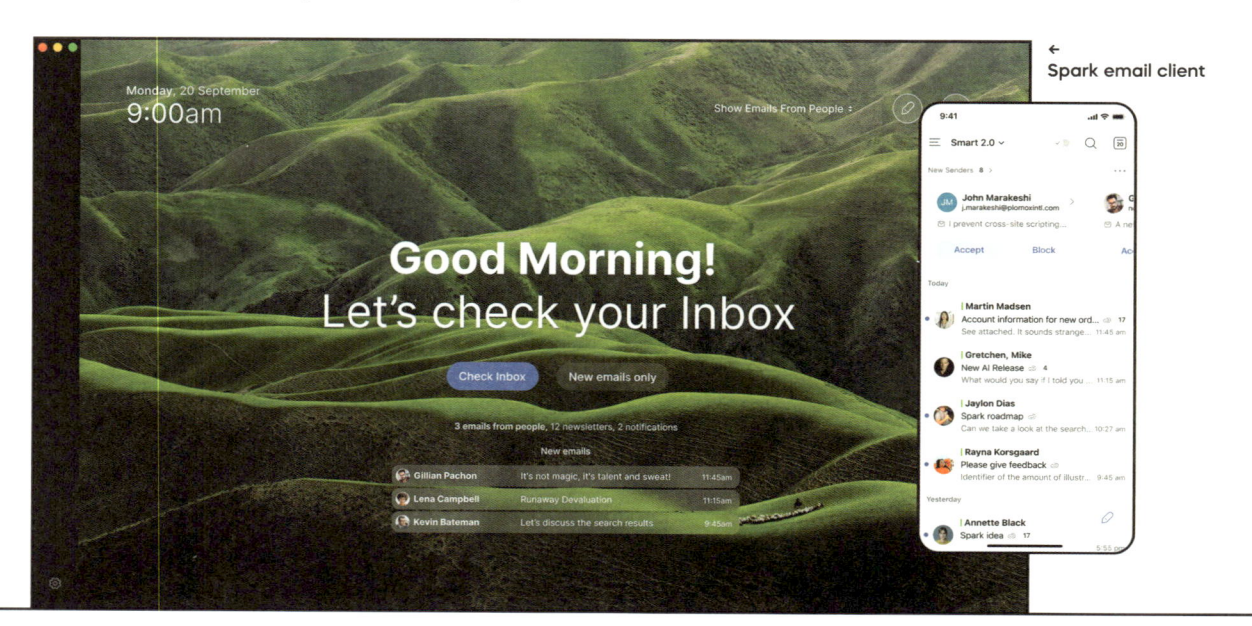

Spark email client

What don't people like about traditional inbox setup. Look no further than the sheer number of messages they receive. Often, important emails simply get lost in the shuffle. Secondly, people are bombarded with unwanted emails and spam. Spark was the first to solve these problems by automatically sorting incoming messages into three categories: Personal, Notifications, and Newsletters. This way, important messages from friends, family, and work are always at the top.

But Spark does more than tidy up inboxes. It creates a modern communication tool for teams and individuals. In today's world, we've grown accustomed to using several different forms of communication — texts, group chats, phone calls — for a single conversation. You may start an exchange via email, then continue it on Slack or Skype. But why? It's because for correspondence today, traditional email doesn't fully meet all of our needs.

In certain cases, it's just easier to use other programs. For instance, sometimes you need an immediate answer about something you've discussed with a colleague or client via email. In this case, an email simply won't cut it because emails aren't expected to be answered immediately. Instead, you need a messenger. That's where Spark saves the day again, with private comments and messaging on top of email capabilities.

In May 2023, Spark +AI launched, making the app one of the first main email providers to deliver AI capabilities. Spark +AI marks a new era in email productivity, helping users craft the perfect email in seconds, and increasing their confidence.

> *"Spark was built to address the epidemic of email burnout. We know that millions of people see email as an unpleasant place to be, a black hole that sucks their time and energy. We help people learn to love their email again and transform it into a real productivity tool."*
>
> — Alex Tyagulsky, Chief Product Officer and Co-founder of Readdle

Simply put, Spark is an email client unlike others on the market. Readdle claims they're reinventing email. Only time will tell whether they will achieve this goal.

Meanwhile, PDF Expert maintains the company's hallmark qualities: convenience and ease of use. PDF Expert is so easy to use, in fact, that it has led to a minirevolution in how we work with PDFs. The app allows users to interact with PDFs in the same way they would with an editable document. Changes to contracts, document corrections, and content updates can be implemented instantly. PDF Expert has made it possible to edit basically anything in a PDF. The product has enjoyed rave reviews as an affordable PDF editor from tech gatekeepers like Macworld. Thirty million downloads later, it looks like PDF Expert users completely agree.

So, how was Readdle able to achieve such a resounding success? Firstly, not all small startups are as ambitious as Readdle. The team set some big goals and hit the ground running. Readdle made it their mission statement to improve people's lives through technology. All of Readle's products increase productivity and cut down on routine tasks so they can spend more time doing what makes them happy.

"One morning we woke up and saw some mind-blowing figures on our computer screens — an 800% increase in sales over the course of one day. We started looking for the reason behind such growth and found it on the pages of the Wall Street Journal. *It turned out that Walt Mossberg had mentioned us in an article. That's when I understood one very important thing — we had to gain coverage in the world's top media to increase the company's revenue."*
— Denys Zhadanov

Secondly, Readdle knew how to get its name out. Readdle board member and former VP of Marketing Denys Zhadanov has played a key role in developing key contacts with the media. It's no coincidence that he made the Forbes 30 under 30 list in 2018. Denys sees his experience as a positive example for young entrepreneurs "Considering Russia's war against Ukraine, we have to show young people that you can start your own business, study at the best schools, follow your dreams, and build a democratic society."

↓
Calendars by Readdle and Documents

169

Grammarly: Improving Lives by Improving Communication

← Grammarly co-founders: Max Lytvyn, Alex Shevchenko, Dmytro Lider

One of the world's leading AI writing assistants did not emerge from Great Britain, the United States, or Australia, but from Ukraine. It's called Grammarly, and it's been in business for over a decade. Its founders, businessmen and programmers Max Lytvyn, Alex Shevchenko, and Dmytro Lider, are neither linguists nor native English speakers. They landed on the idea for Grammarly after stumbling on an insight from a different corner of the writing space.

"It was difficult to teach computers to help people communicate better. We were frequently told that it's virtually impossible."
— *Max Lytvyn, Alex Shevchenko, and Dmytro Lider, Founders of Grammarly*

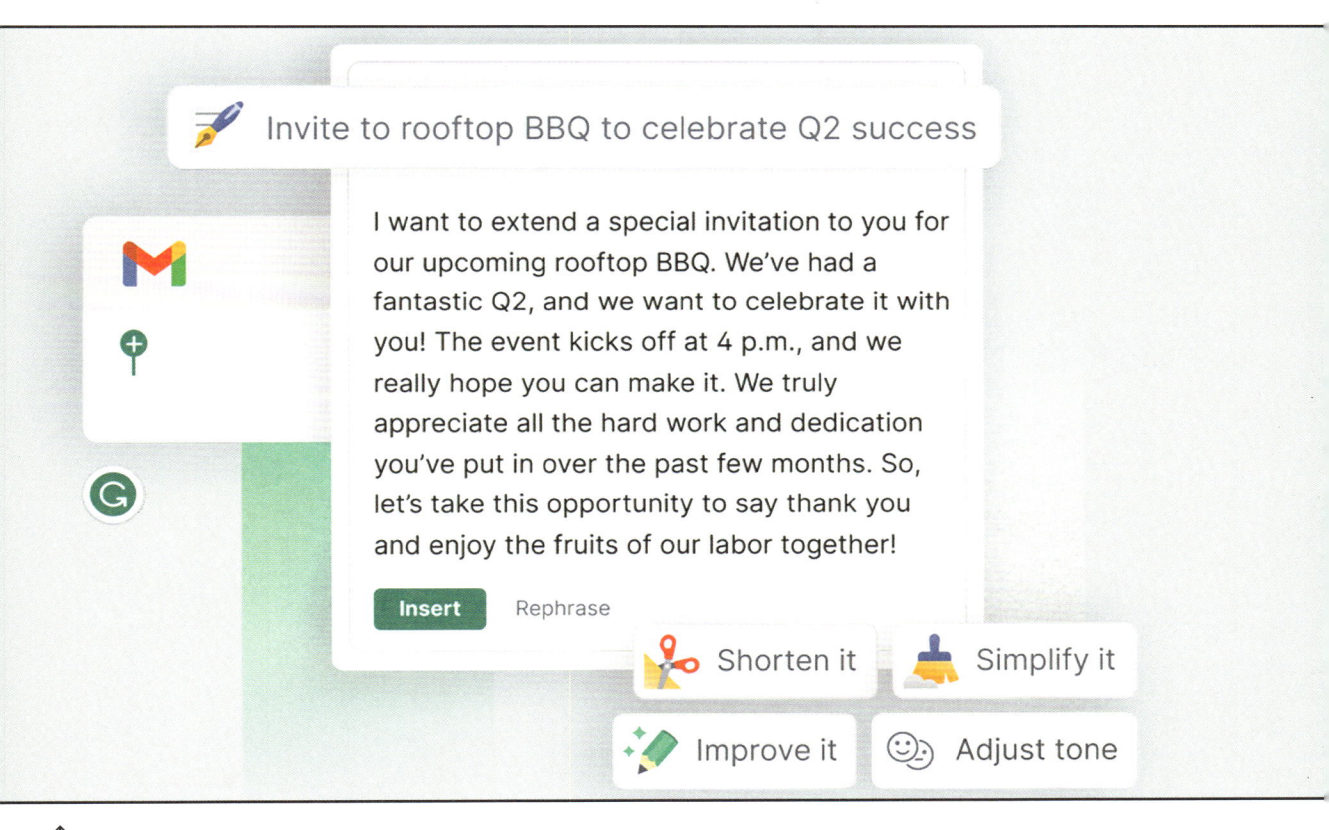

↑
Grammarly's generative AI features

"We realized that most people plagiarized texts because they found it too difficult to write by themselves," the Grammarly trio says. The idea for an automated plagiarism checker was born. They found that people often view their communication skills as static, and there were no technology solutions on the market that improved these skills. Their realization ultimately inspired them to develop the app that today helps helps over 30 million people and 70,000 professional teams ideate, compose, revise, and comprehend texts in English every day. The company's mission statement says it all: "to improve lives by improving communication."

After selling their initial plagiarism checker product to to Blackboard, the world's largest education technology and services company, Grammarly's founders targeted a much larger audience for their new product. For a long while, they bootstrapped the company. Max, Alex, and Dmytro focused on building the product themselves with a small team. They decided not to involve investors at this point. This is quite unusual for a startup in the software space, where projects typically seek investment as quickly as possible. The Grammarly trio, however, tried to avoid partnering with anyone for as long as possible to keep the decision-making power in their own hands. Instead, they focused on developing their product. Their plan? Build up a strong user base first and then grow profits from there.

The founders of Grammarly embarked on an ambitious mission when they began the project: coding an online asistant that would provide instant feedback and recommendations on how to write clearly, effectively, and confidently. With a smart assistant to help refine form and tone, writers would be able to focus on the content of their messages. They decided to build their service for English. As there are over two billion English speakers worldwide, the app's positive impact could be truly global.

← Grammarly's Kyiv hub

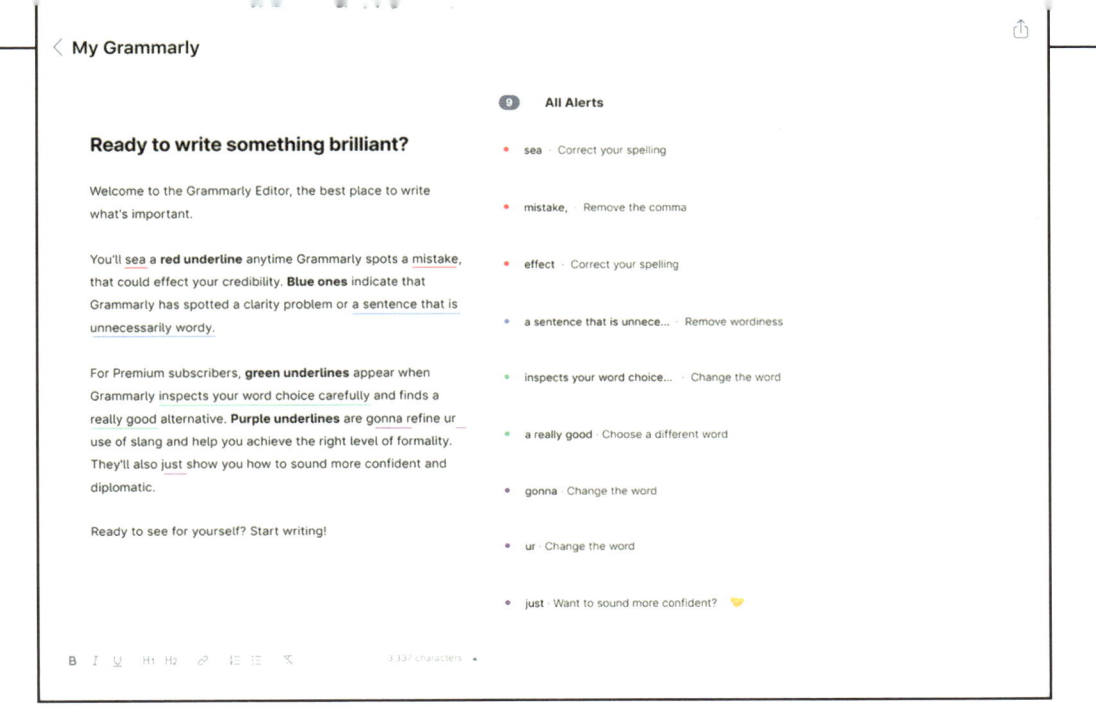

Grammarly for iPad

Grammarly quickly went far beyond grammar and started to offer comprehensive writing assistance, including full-sentence rewrites, tone adjustments, and fluency suggestions. Today, the company uses a combination of advanced machine learning with human expertise to offer on-demand communication support to individuals and enterprises, whether they are starting from scratch or revising an existing piece of writing.

"We are all fluent in English. At the same time, as non-native speakers, we understand how it feels to learn English from scratch. Grammarly is used mostly by native English speakers, but we've integrated a learning component into Grammarly that helps everyone to become more effective communicators," Grammarly's founders say.

What does the technical side of Grammarly look like? The company uses a variety of innovative approaches — including generative AI, natural language processing, advanced machine learning, and deep learning — to consistently break new ground in its product offeings. Grammarly's newest generative AI features help people quickly compose text, rewrite paragraphs, draft replies to emails, and more with a prompt. They are aware and account for personal voice, offering relevant and personalized suggestions that respect user agency and authenticity.

> *"Grammarly's product strategy follows a core principle: We want to augment human potential, not reduce personal autonomy.*
> *We believe technology should be built to solve real user problems, enhance understanding and productivity, and empower human connection and belonging. AI is successful when it augments humans — their intelligence, strengths, and desire to be understood,"* says Grammarly's CEO Rahul Roy-Chowdhury.

Grammarly's distinct strength is that the product offering is platform-agnostic — it works right where people type across web, desktop, and phone, including 500,000 apps and websites. The team follows the timeless commandment of Dieter Rams, the genius industrial designer behind Braun: good design is as little design as possible.

From its rollout in 2009 until 2017, Grammarly continued as an independent, self-funded startup. Its founders decided to bring on outside investment only after eight years, at which point they raised $110 million in their first investment round. Their focus was on finding the right investor group with a proven track record of building leading consumer internet and deep technology companies. They found what they were looking for in a group led by General Catalyst, with participation from IVP and Spark Capital. In 2019, Grammarly received another $90 million from General Catalyst, bringing the company's valuation to over $1 billion. And with that, Grammarly became the first Ukrainian unicorn.

→
Grammarly's Tone Detector

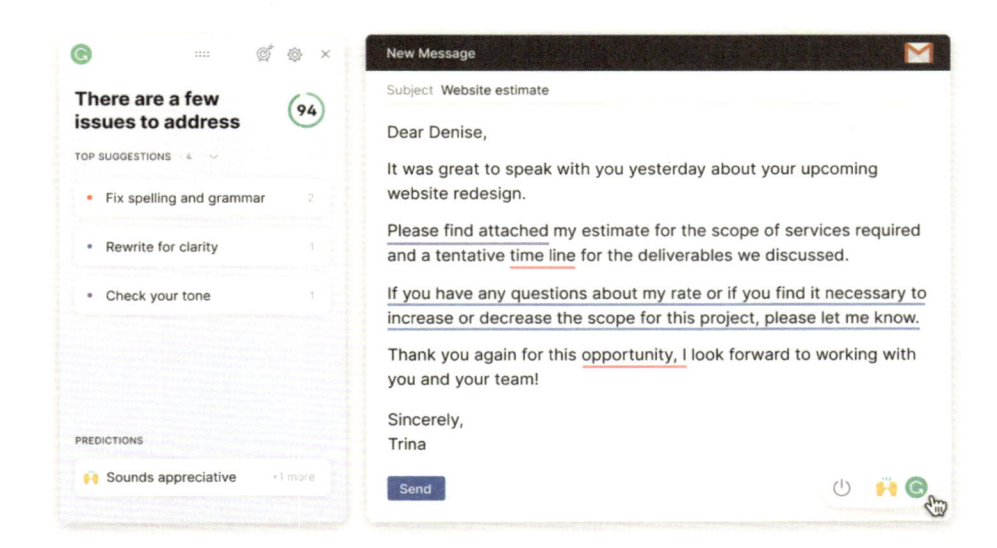

There are a few issues to address (94)

TOP SUGGESTIONS 4

- Fix spelling and grammar — 2
- Rewrite for clarity — 1
- Check your tone — 1

PREDICTIONS

Sounds appreciative +1 more

New Message

Subject: Website estimate

Dear Denise,

It was great to speak with you yesterday about your upcoming website redesign.

Please find attached my estimate for the scope of services required and a tentative time line for the deliverables we discussed.

If you have any questions about my rate or if you find it necessary to increase or decrease the scope for this project, please let me know.

Thank you again for this opportunity, I look forward to working with you and your team!

Sincerely,
Trina

Send

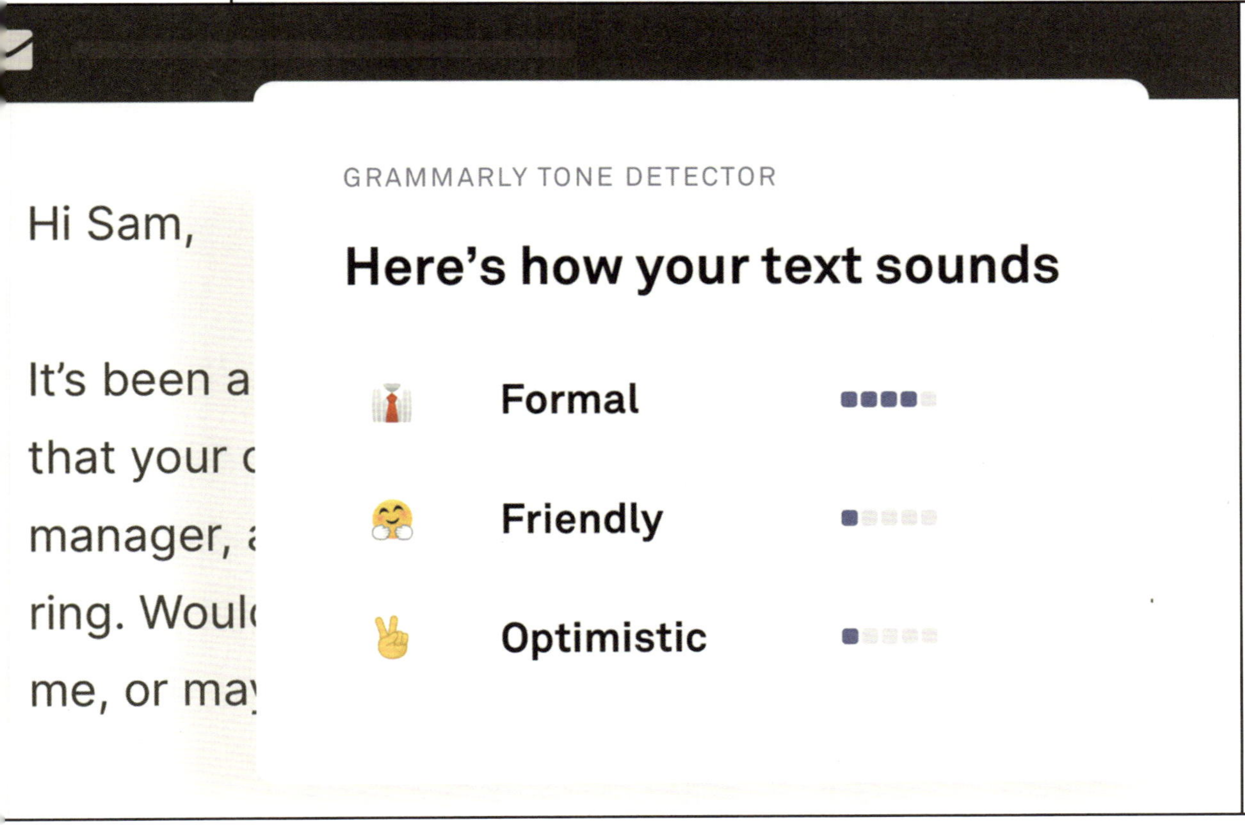

Hi Sam,

It's been a
that your
manager,
ring. Woul
me, or may

GRAMMARLY TONE DETECTOR

Here's how your text sounds

👔	Formal	■■■■□
☺️	Friendly	■□□□□
✌️	Optimistic	■□□□□

175

The company's growth didn't stop there. Since Grammarly's initial bid for investments, it has raised a further $200 million from investors including Baillie Gifford, the UK investment management firm and funds and accounts managed by BlackRock. Grammarly's market penetration has also scaled along with its funding. In April 2023, the company launched new generative AI features that accelerate productivity where people write. Combined with Grammarly's advanced writing support, this functionality helps to generate text instantly, enabling people and businesses to unlock their potential, save time, and get more done — while prioritizing trust, security, and authenticity. One of the company's current priorities is to deliver support for the AI-connected workspace and help businesses to better access and use information across their organizations, elevate teams' skills and productivity, and create content with increased speed, relevancy, and comprehension.

Grammarly's corporate culture also contributes to its success. All team members know the company's "EAGER" values: ethical, adaptable, gritty, empathetic, and remarkable. They inform how the company approaches every process within the organization — from developing a product that connects people to supporting its users with empathy to creating an inclusive workplace for the global team.

Grammarly has long helped people and teams communicate more effectively with suggestions that make their writing clearer, more concise, and more compelling. With generative AI, the company is moving beyond the editing stage to support its users across the entire communication lifecycle, including conception and composition, and further deliver on its mission of improving lives by improving communication.

← Grammarly's Kyiv hub

Athena, **a support manager**, helping a client resolve a shipment issue.

Please rate our QoS with this short survey.

ACMECO STYLE GUIDE

~~QoS~~ → quality of service

Most customers are unfamiliar with this abbreviation.

Grammarly in Figures (as of January 2024):

- 70,000+ professional teams served through Grammarly Business
- 30+ million daily active users
- 1000+ team members across Europe and North America
- 5 hubs in Kyiv, Berlin, San Francisco, New York, and Vancouver

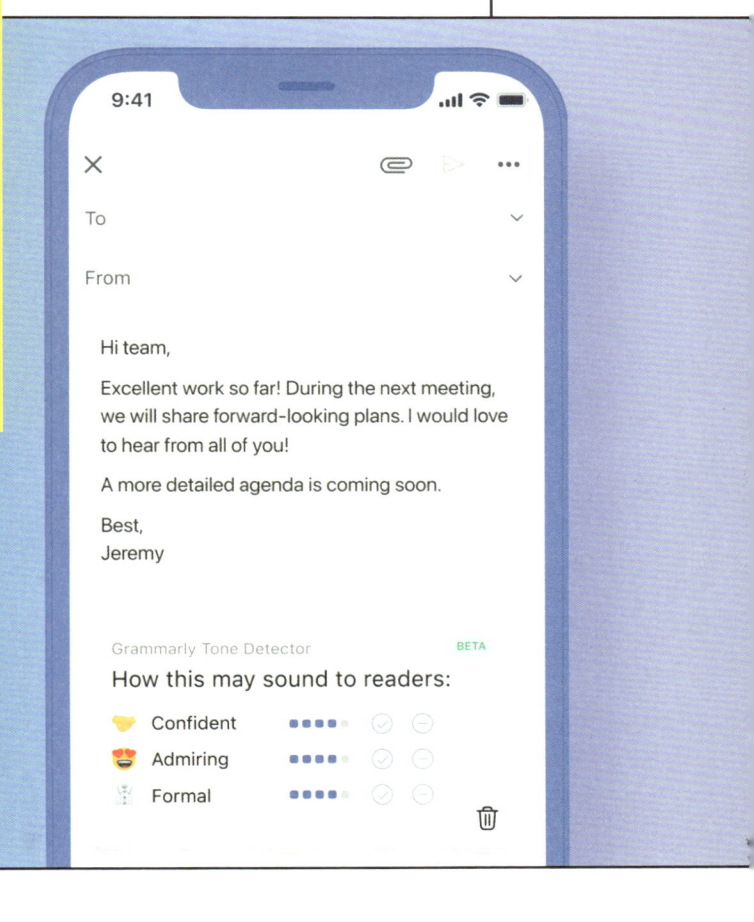

↗
Grammarly
Style Guides

Depositphotos: Slow and Steady Doesn't Win the Race

←
Dmytro Sergeev, founder of the Depositphotos content platform

The success of Ukrainian content platform Depositphotos stands out as a true David and Goliath story. On one side there was a small startup out of Kyiv that used ingenuity to fight its way into a competitive marketplace. On the other, there were giants like Shutterstock and Getty that, initially, claimed the entire niche. The reputations of these companies were so strong that in business circles, it was considered a bad idea to enter commercial photography as a new player.

"Millions of people dream about monetizing their creativity. Why not give them a chance? And why not turn it into some grand enterprise? We've got all the know-how we need to do that."

— Dmytro Sergeev, Founder

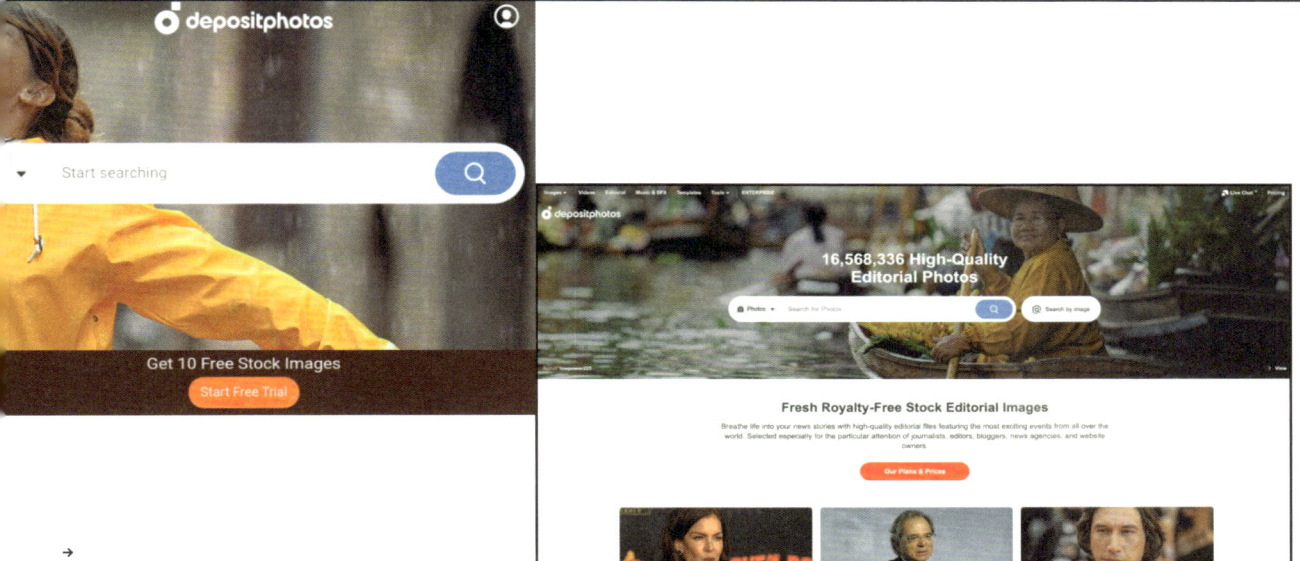

→ Depositphotos home page. Image search.

In 2009, Dmytro Sergeev, Depositphotos founder, dared to challenge the monopolists with his own project. His bold decision to try to break into the global market was not typical for Ukraine, but it turned out to be a success. In a matter of weeks, Depositphotos had put up a whopping one million stock images for sale. How was that even possible? Well, several factors were at play. Almost 60% of contributors lived in former Soviet republics. Depositphotos was the first ready-made platform for talented photographers in the region to bring their content to the international creative market. Beyond the opportunity for exposure, Depositphotos' excellent content managers and its personal approach earned the company trust in the creative community.

But there was another crucial factor: Dmytro paid one dollar as an advance to anyone who uploaded a stock photo. In the wake of the global financial crisis, it's no wonder that he soon had hundreds of amateurs lined up.

← Depositphotos office in Kyiv

He explains that this strategy gave the company an important boost in the early days: "We believed that any means were good to break into the market. We got caught up in the hype... this way, we could quickly beef up our pool of collaborating photographers and show them that we were serious about doing business." That strategy didn't last long, but it did the trick. Eventually, the company won a three-percent share of the global stock photography industry. Dmytro says that throughout this initial push, he never lost sleep over their competitor.

> **"We didn't really think about it. We didn't do comprehensive market research or brace for a blitz campaign. We were just looking for a model that would build upon our previous experience."**
> — Dmytro Sergeev

He chalks the company's success up to the genius of the people behind it. "A few talented guys who used to contribute to the microstock industry joined us," he remembers. "I was very lucky to have such a rock-star team." Why did they go with a microstock? The fact that Dmytro's wife was a photo buff was one of the reasons he chose photos specifically and not some other content project.

At the same time, Dmytro and his team had a good deal of experience already in content sharing and had recently sold a previous project to the tune of several million dollars. The founder says that these funds, coupled with their niche expertise, helped kickstart Depositphotos. "We invested roughly a million dollars into this startup until we made sure that it could really take off. We spent that money on developers, marketing, servers, and advance payments to photographers. Later, investors took over."

"What would you have done if you hadn't started Depositphotos?"

> *"I don't know. Something else. I don't really care what I do as long as it gets me hyped up."*

→ Depositphotos home page. Sounds search.

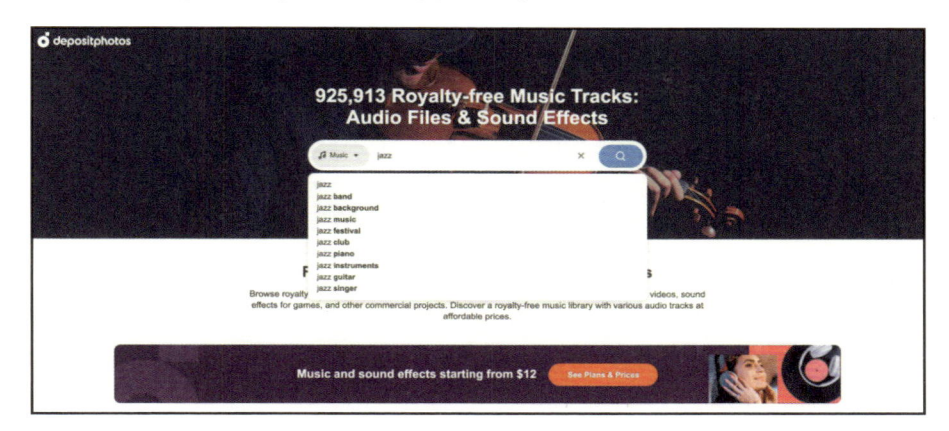

Depositphotos today is much bigger than the scrappy startup it started out as, but it is still smaller than many of its competitors. The small size keeps the business agile. In order to constantly and effectively grow as a marketplace, the company began experimenting with neural networks in 2018. The outcome of their efforts was a set of AI-based tools to help the team, contributors, and users make the most out of their collaboration with Depositphotos.

To keep the platform accessible for its growing global audience, Depositphotos now features photo descriptions in 29 languages. And it has introduced advanced search filters that allow users to search by image, color palettes, season, time of day, location, orientation, number of people, contributors, and more. In addition, smart suggestion algorithms now help users discover more content options and quickly find what they're looking for.

The platform boasts a number of advantages that help them attract premium contributors and earn user loyalty. Chief among them is user experience, reflected in its greater flexibility and support. Depositphotos prides itself on its 24/7 support for contributors and users. The price barrier is also low, while the quality bar remains as high as its competitors. On the user side, Depositphotos is able to satisfy even the most specific requests in-house or via its network of content creators. These factors keep bringing people back to the platform time and time again.

When it comes to content, Depositphotos has found that interest in particular categories remains strong from year to year. The takeaway? Any content platform set up to cover people, food, family, business, and icons will find a market. Beyond those standard categories, seasonal themes are also high-demand, as well as trends like candid photos. But when users' options are almost unlimited, they become more and more demanding. For this reason, Depositphotos has shifted its focus from quantity. Now, it is all about quality.

With thousands of photo submissions, how does the team moderate quality? A neural network-based approach to product development gives the company a major competitive advantage. They use artificial intelligence to filter this vast quantity of submissions, freeing up resources and talent for strategic work. Before looping AI into process, it took around a week to manually add file information from photos to each appropriate database. Today, it takes as little as five minutes.

And it's not just speed where Depositphotos wins. Artificial intelligence also helps the team detect image duplicates and banned or misspelled keywords. This ensures that all the submissions are safe and high-quality. At the same time, the platform keeps copyrights secure via blockchain technology.

Contributors might wonder about their chances of having one of their photos selected for use. Dmytro says that the most important factor is, as always, quality. "Take a dozen photos for starters, but put your heart into them." This is good advice, as Depositphotos rejects roughly 20% of submissions. To make it easier for would-be contributors to make the cut, the company guides content creators through common questions, addressing specific niches and styles in the industry. Depositphotos also publishes an annual Creative Trends report to ensure that submitting content is a win-win for the creator, the company, and the user.

Depositphotos' success over the last decade has enabled the company to support and inspire creatives through a series of other successful projects and features, including graphic design tool Vista Create (Crello), complete with the AI-powered Image and Video Background Removers, Image Upscaler, and Reverse Image Search tools. These latest innovations reduce tedious technical work to the click of a button, giving creators back their time to focus on creativity.

← One of the working days in the Kyiv office

→ Image search by keyword

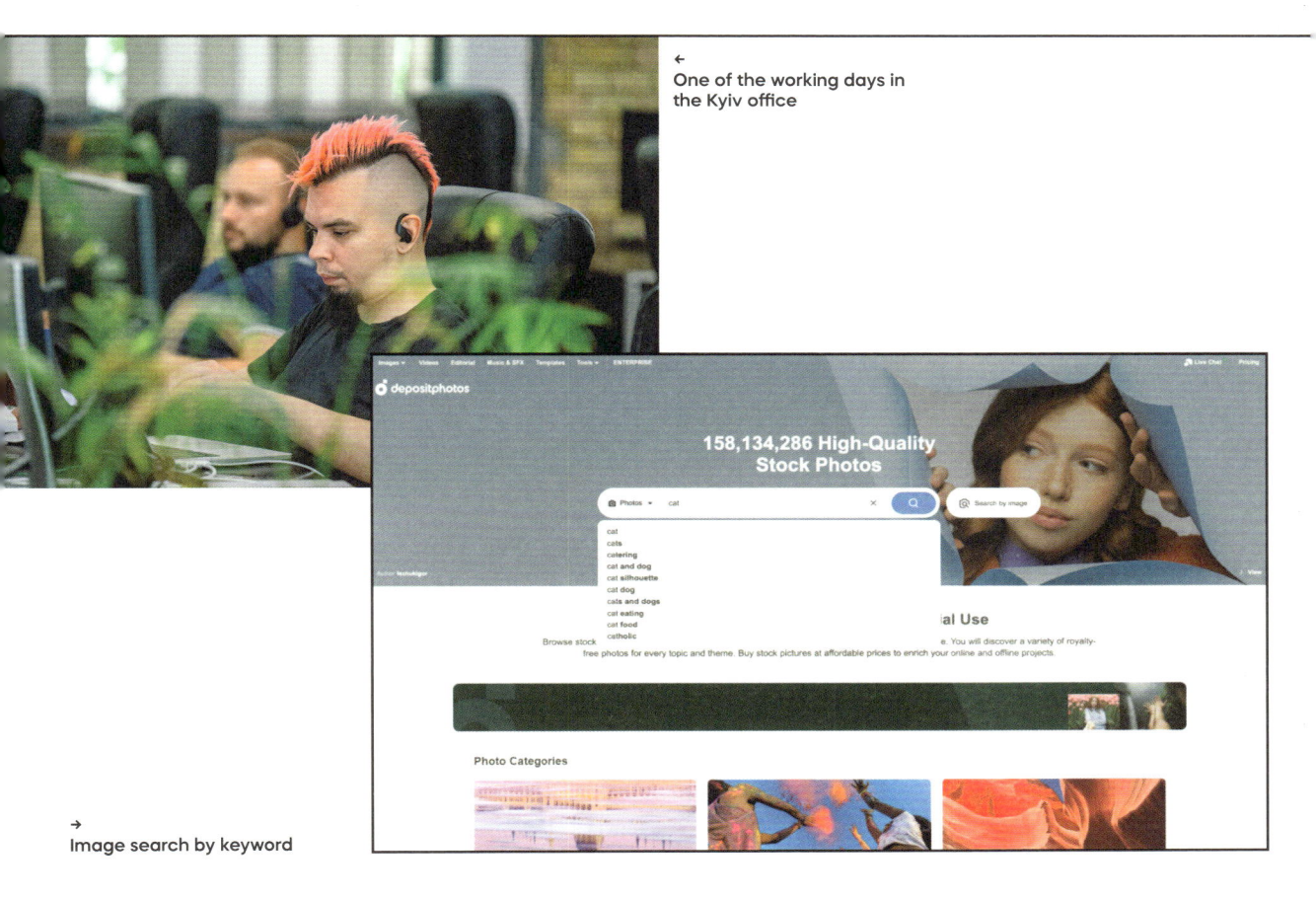

The company's vision doesn't stop with digital services. Depositphotos' photo studio is one of the largest providers of photo and video content in Europe, creating on-demand and stock content for clients across the globe. This project shares Depositphotos' values of quality content and excellent customer service, complementing each other within the same creative ecosystem.

Today, the Depositphotos library consists of over 270+ million files, making it the one of the largest stock library on the market. The company collaborates with 100,000 talented contributors, who upload five million new files a month. The marketplace features a vast selection of photos, vectors, illustrations, videos, sound effects and music tracks for every taste and budget. It doubled its size in just two years, and was acquired in 2021 by global design company VistaPrint. Through all of this growth, Depositphotos' goal remains the same: to provide users with a rewarding experience at every stage of their interaction with the platform.

Petcube: Keeping Your Pet Happy When You're Away

← The company was founded in 2012, in Kyiv, by Yaroslav Azhnyuk, Alex Neskin, and Andriy Klen.

Pet owners, primarily cat and dog owners, don't like being away from their furry pals for too long. But what if you're at work all day? Petcube is a fun little device for pet owners that means they never have to leave their animals completely alone. It has a built-in wide-angle camera, speaker, microphone, and interactive features to entertain or reward dogs and cats. This gadget allows pet parents to keep an eye on their pets remotely, as well as talk to, play with, or even treat them. You can also take pictures of your pets and instantly post them online.

"How's Spot doing?" Animal lovers are willing to pay $200 or more — that's how much the original Petcube model had cost — to get an answer to that question. The numerous investors who have backed Petcube's team knew that too. The project has raised over $25 million dollars over the course of multiple rounds of funding. Quite a lot for a pet cam, wouldn't you say? First off, it's a top-quality product with an award-winning design. It's obvious who Petcube is intended for, and it's obvious how it works. The video on the company's website provides an excellent visual without tedious explanations, all within two or three minutes. Oftentimes, simplicity is the key to commercial success. At the same time, the pet industry is an enormous market.

Petcube Bites 2 Lite interactive pet camera lets owners to monitor and treat their furry companions on-the-go.

When you look at Petcube, you realize that it isn't a completely new product. Stores began selling web cameras well before Petcube came along, and all of them, though poorly, can perform the main function most pet owners need: showing them what their pets are up to. The market differentiation comes from the improved user experience and the extra fun features. The laser pointer and the treat dispenser provide added benefits, giving pet lovers ways to interact and care for their four-legged family members when they can't be there in person. These additional features have clearly positioned the product on the market and have even shaped the market for other "connected pet" devices.

Many popular magazines and websites wrote about Petcube once it came out, but stellar customer reviews and word of mouth wound up being the best form of promotion. Even celebrities have gotten in on the hype. One of them is *Harry Potter* franchise actress Emma Watson, who shared with her fans, "Oh my god, and Petcube! This is life-changing."

The hundreds of thousands of people who have purchased Petcube use it, on average, eight times a week for more than 50 minutes a session. Think of how many users spend more time chatting with their pets than with their parents!

Petcube's success proves that, although the market may seem oversaturated with various devices, go-getters can always seek out an opportunity to create something new. However, the product has to be intuitive and user-friendly, which Petcube very much is. That's no accident. The Petcube team understands that designing a product is a process, more like sharpening a sword than a one-and-done achievement.

The Petcube Play 2 camera with a built-in laser toy to see, talk, and play with pets remotely paved the way for the connected pet consumer category.
↓

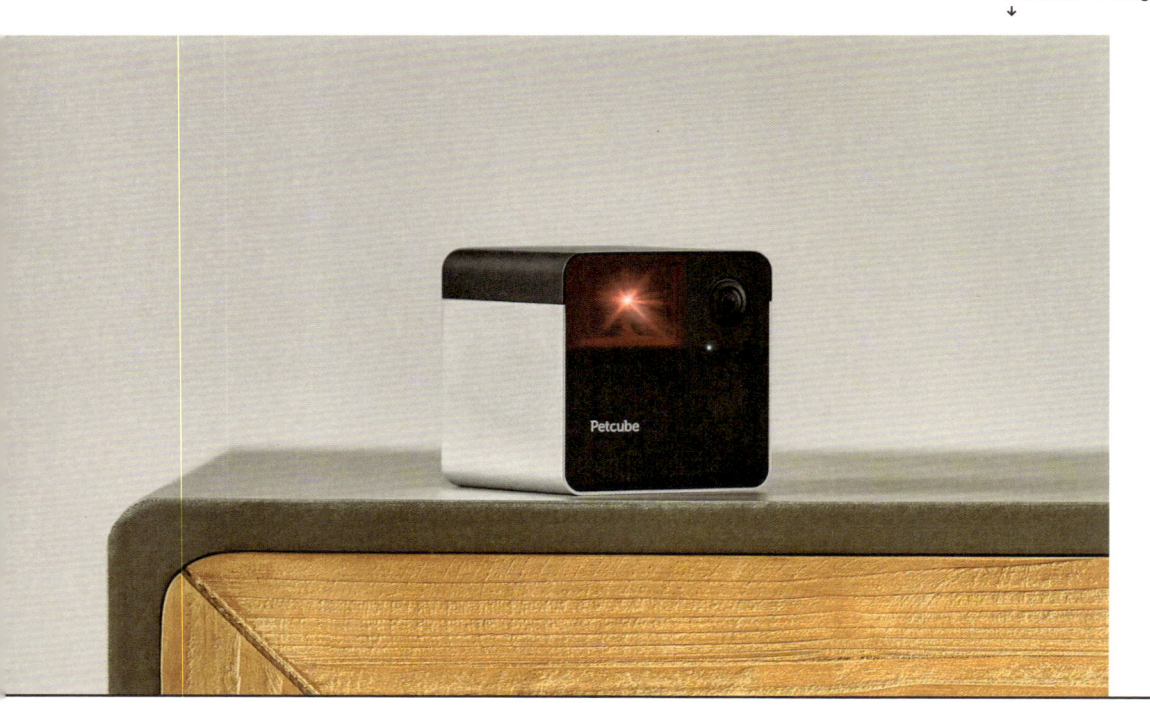

They also understood the irrationality of working on a gadget for several years only to discover that no one needs it. The founders decided to reign in their costs and try to ascertain potential demand by launching a Kickstarter campaign. Their thinking was that it was better to work on a project that had some real interest behind it, even if it ended up not working out in the end. Petcube's first Kickstarter campaign was a resounding success, generating 1,700 pre-orders for a total of $250,000.

Yaroslav Azhnyuk, a co-founder of Petcube, has put together a few recommendations for those thinking about launching their first Kickstarter campaign. First off, study your predecessors' experiences. Secondly, don't just blatantly copy other people's ideas. Try to answer a few fundamental questions: Who is my target audience? What value does my product offer? How can I convey that to potential buyers? How can I check whether or not people understand what I'm offering?

Naturally, the choice of main market is crucial. Launching products quickly without spending an arm and a leg just isn't an option. Moreover, the purchasing power of potential users has to be considered. Like most other successful projects, Petcube was geared toward the Western market. According to Yaroslav, they were focused on the U.S. from the very beginning, although Petcube is also sold internationally. In spring 2023, Petcube announced the start of sales of its products in Ukraine.

Initially, Yaroslav and his partners covered the costs of developing their product and launching a Kickstarter campaign themselves (roughly $40,000). But, naturally, they needed a much more sizable investment to grow their company. Yaroslav handled fundraising. In his experience, talking with investors is itself a full-time job that requires skill and patience.

> **Thinking back on what he learned in the U.S., Yaroslav recalls, "I did fundraising in the U.S. More precisely, I learned how to fundraise there. Basically, I tried to figure out how the market works. Our next source of funding came from AVentures. Subsequently, it became our go-to firm for fundraising. As of Petcube's series A in 2017, about 80% of $14M raised came as referrals from AVentures, although the firm itself has only invested $1.5M. I can't say that we had a flood of investors. We pinpointed a few. Essentially, seeking investment has been a full-time job for me over the past four to five years."**

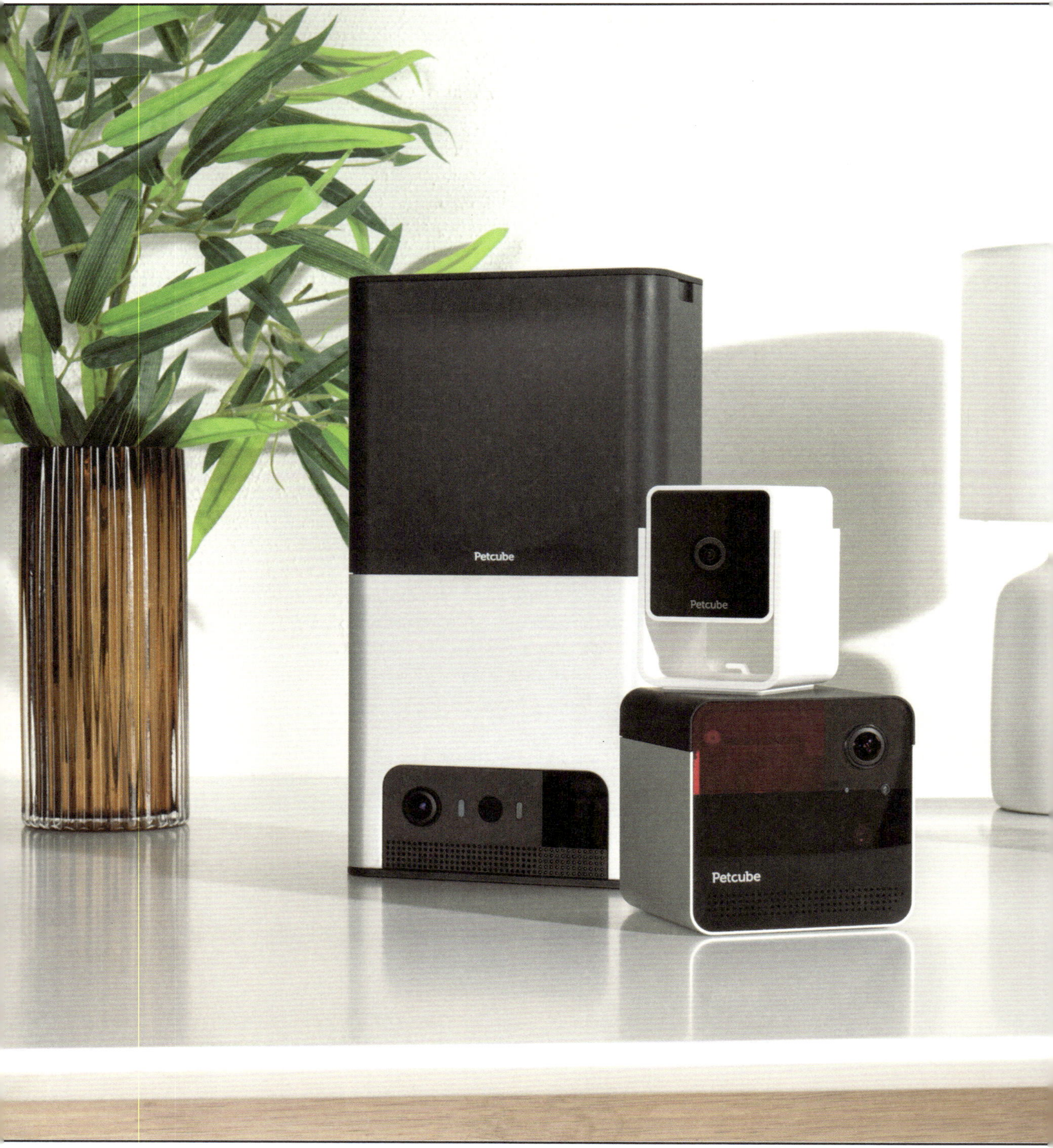
Petcube
Petcube
Petcube

The real Petcube runs on a Linux computer. The company partnered with another Ukraine-based developer Obreey Products, the maker behind PocketBook e-readers. They provided a board, which the Petcube team reconfigured to meet their needs.

And then Amazon came along. "Sometimes, Amazon comes to you, like they did with us. It's very easy to get started with Amazon. There are a few different options — you can set yourself up on Seller Central. As of 2019, Amazon is a Wild West, sort of like Google was in 2005, and there's plenty of opportunity for growth."

Yaroslav Azhnyuk on Petcube's future:
"The company's mission is to connect pets to the internet, and give them a voice. Of course, that seemed like a mere fantasy at first — hooking pets up to the internet, giving them all the devices that would enable us to talk to them at any moment, seeing them whenever, sharing our pets with our friends so they could talk to them."

←
Petcube interactive cameras are made of high-quality hard plastic and are entirely pet-proof.

Anyone who talks with Yaroslav can't help coming away with two lessons on what it takes to build a successful Ukrainian company: don't be afraid of your competition, and create new market niches — that way you won't have to compete against anyone. That's not very easy, but it's easier than many think. Yaroslav says that the more users interact with their device, the better the machine learning algorithms get.

"We can start to understand our pets better, see behavioral patterns, figure out when our pets aren't feeling well, and understand what they want. I think that this incredible, fascinating dream about animals talking will become reality in the next 5–10 years."

Ajax Systems: An Idea Is Nothing, Execution Is Everything

←
Oleksandr Konotopskyi,
Chairman of the Supervisory
Board of Ajax Systems

How are deadly fires prevented in one of the most isolated locations on Earth? The Vernadsky Research Base is home to a small group of committed researchers who study the Earth's magnetic field and measure climate change. In this extreme environment where hurricane-force winds blow most of the time, a small flame can turn into an inferno in a matter of minutes.

In fact, that is exactly what happened in 2020, when the limits of the dated fire alarm system at Russia's "Mirny" station were unexpectedly tested, with disastrous consequences. After an undetected fire razed vital portions of the base, the Vernadsky team took their cue. They moved to upgrade their own fire detection equipment to the Ukrainian-made system trusted around the world: Ajax.

↑
StreetSiren by Ajax Systems

Ajax is a professional smart security system with a product portfolio that includes more than 135 devices — detectors of fire, flood, motion, door opening, and window breaking, to name just a few — and a hub controlling this system. The wireless devices communicate using Jeweller, a radio protocol proprietary to Ajax, while the hub works on a real-time operating system, OS Malevich. The company has recently rolled out Fibra, a new line of wired devices. Just like the Jeweller-based devices, they are notable for their anti-sabotage features, as well as hassle-free installation and maintenance.

Ajax is produced by Ajax Systems, a Kyiv-based company established in 2011 by then-25-year-old Oleksandr Konotopskyi. The equipment is manufactured locally in Ukraine and Turkey at the company's two factories, employing over 1,800 people.

Its products are sold in 169 countries across the world and the company has been valued by *Forbes* at more than $500 million dollars. The founders value their brainchild even more.

"Even if someone offered me $1 billion for Ajax, I'd say 'no,'" says Oleksandr Konotopskyi.

What is the company's target market segment? Who's buying it? Among Ajax's more than 2 million users, one can find all kinds of clients, from typical homeowners to the Vernadsky Research Base in the far reaches of Antarctica and everyone in between. Ajax systems are fully reliable, scalable, and satisfy the most rigorous demands, as proven by Ajax's extensive certifications and awards. All products are certified according to strict European ISO/IEC quality standards, as well as the ISO Environmental Management Systems standard, which certifies the product's low environmental impact in production and use.

A batch of Ajax DoorProtect devices is being transported.

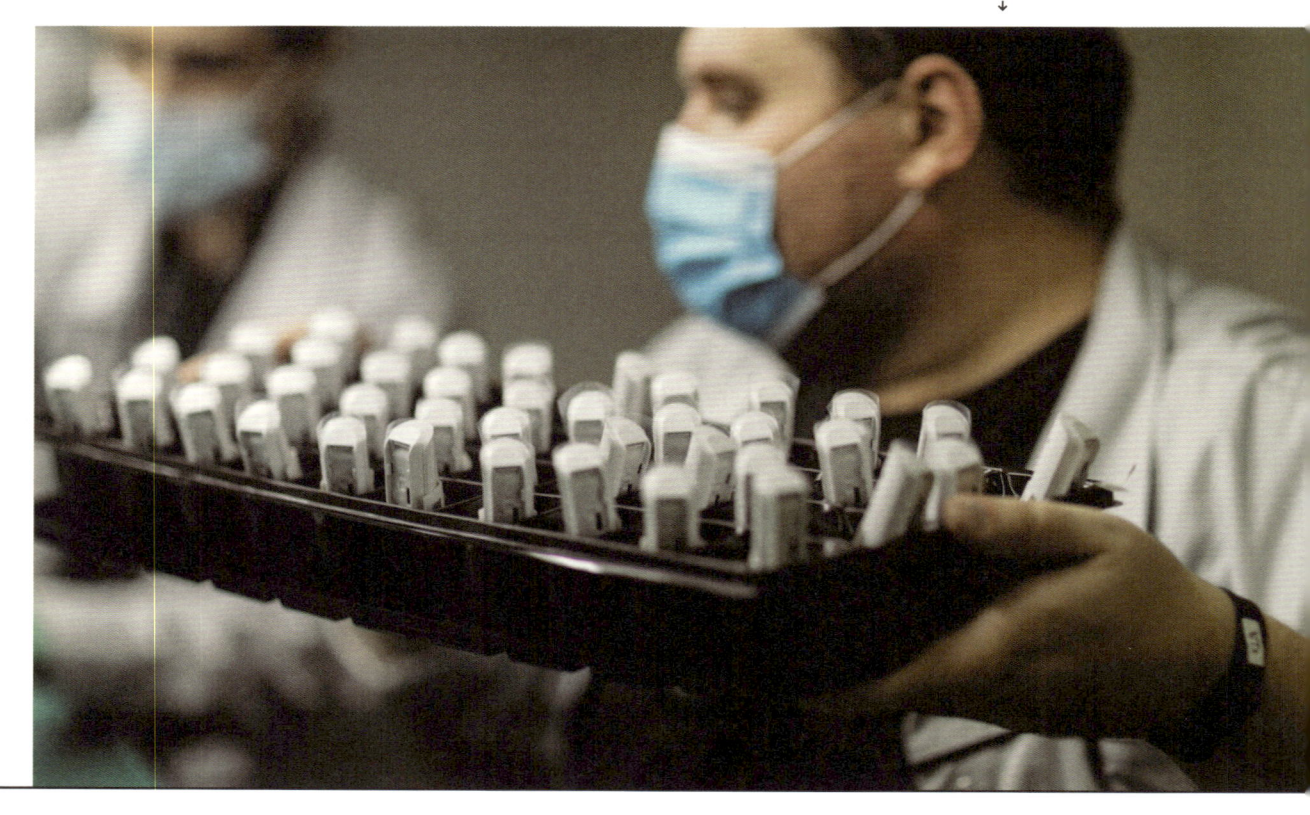

At the same time, Ajax systems are affordably priced. Ajax StarterKit, a basic home kit, costs a little over $300.

Installation is fast and straightforward through a mobile app, but inexperienced users will likely benefit from the guidance of a professional. A detector set up in the wrong place might either generate false alarms or malfunction just when it's needed — and it wouldn't be the fault of the system. For this reason, the producer sells most of its devices through professional companies with expertise in this area. In fact, installation, is where the product delivers additional value, albeit value that cannot be measured in numbers. It's the fun factor. Most people don't ask for much from a security system, beyond that it does its job. Ajax, on the other hand, approaches its products with the same user experience-centric philosophy behind consumer products like phones or computers. Starting from the moment the end user takes the system out of the box, user experience is front and center. Ajax might well be the only "fun" alarm system in the world. The company prides itself on making security systems that are as enjoyable to use as the latest-generation smartphone.

As it's often the case, Ajax started with a coincidence, and for the first couple of years, it followed quite a cliché path.

"I started my business as a distributor back when I was a student at Kyiv Polytechnic Institute. At the time, it was not obvious yet that IT was a rising star of our age. Back then, a good business was to buy some goods in China, ship them to Ukraine, add a high markup — at least 200% or maybe even 300% — and then sell them." explains Oleksandr Konopskyi.

Oleksandr started his own business in 2008 following the familiar model: buy cheap from China and sell at a profit in Ukraine. Those initial home security systems sold well. In the wake of the financial crisis, Ukrainians could hardly afford European solutions, and the Chinese imports looked like a good alternative. "For the first three months, everything was going great," Oleksandr says. "But then we started to have some issues with the Chinese-made technologies. Devices exploded. People had problems with alarms. Our business spiraled downward. I was very brave but not very smart. So, I decided to launch R&D and develop our own security systems."

That's when Ajax was born. The systems sold quite well, and, encouraged by the demand on the local market, the company's partners decided to go international. In 2013, Ajax brought its security system to a foreign exhibition for the first time. Oleksandr intended the trip to be the company's first step towards international expansion. But everything went terribly wrong. "No one showed any interest in our product," Oleksandr remembers. Ajax may have stayed in the shadows, but Oleksandr had a realization that would define the future of the company.

> "When you worked on the local market, you had your sales channels in place and it was easy to do business. But when you entered foreign markets with their own producers and distributors, your product had to be better. You just couldn't afford to be only 30% better than your competitors. Your product had to be king."

The early Ajax was obviously no king. For many, if not most, entrepreneurs such a failure would be their first and last attempt, but the Ajax team didn't let that stop them. They got down to creating a super product that would outshine all their competitors.

The first step? Starting from scratch. The team studied all the weaknesses of their existing system. A three-year R&D process resulted in unique wireless technology patented in the U.S., fresh product design, and revamped hardware and software developed by the company's dedicated in-house team. By 2016, Ajax was holding its own on the international market.

The first signal that they were onto something good came at an industry show in Las Vegas. "We were standing by our exhibition stand, and visitors from across the world — from Venezuela to Australia — asked us: 'Where are you guys from? You got a top-quality system. Are you from Germany?' And we said: 'No, we're not from Germany but from another European country: Ukraine. It's a rising star, a new tech tiger. You'll hear about it pretty soon.'"

Later, Ajax participated in industry shows in Britain and Germany, meeting new partners and distributors and establishing an international sales network. Over the next two years, Ajax was recognized as the best security system in Britain, France, and the UAE, and it became one of the most award-winning security system in Europe.

Ajax manufacturing in Kyiv, 2021

Funding was no small part of the company's success. Ajax received a $1 million grant from SMRK, a venture capital fund that believed in the promising solution. Ajax used this investment to ramp up the product into a fully-functional advanced security system. In 2019, Horizon Capital invested another $10 million in the company. Horizon bought part of the stake in Ajax Systems from the SMRK venture capital fund and the company's C-suite. The long-term goal of this cooperation was to guide Ajax Systems to IPO.

So, what makes Ajax's system that great? Which of its features have attracted numerous clients, and what promise did investors see in it?

According to Oleksandr, the primary advantage of Ajax is user experience. "Ajax is like Apple or Tesla, while our competitors are like a 1991 PC with Windows NT. Put Ajax Systems and competing systems side by side, and you'll know which one is ours right away." Ajax prides itself on quality at every step, starting from the design of its equipment through to the advanced engineering and deep tech involved in its products. This is the driving force behind Ajax's exceptionally high customer loyalty, earning the company a Net Promoter Score of 87.5%. It just shows that professional products can be people-friendly.

Ajax's superior quality has also been recognized with no less than four industry awards between 2017 and 2020, including the 2017 Security Excellence Award (IFSEC International) and the 2020 Intruder Alarm Product of the Year (MotionCam). Most recently, in 2021, *Forbes* ranked Ajax third among high-performing Ukrainian startups.

↓ Boards after dip mounting

But what truly sets Ajax apart is its commitment to innovation. Driven by high corporate demand, Ajax made the decision to turn down ready-made open-source solutions. Instead, it built its products around in-house solutions. The most impressive example is OS Malevich, a proprietary real-time operating system developed by the Ajax team. Protected from viruses and cyber-attacks, it communicates with cloud servers over several channels and updates unassisted "by air."

Another example is Jeweller, Ajax's own radio technology. Jeweller provides a record-breaking wireless communication range of 2,000 m (6,562 ft.) in open space, as well as high energy efficiency. The batteries in Ajax detectors work for up to seven years. Ajax Fibra protocol is yet another example. With its AES encryption algorithm, authentication for every communication session, and other anti-sabotage measures, Fibra takes wired security to a new level, giving the company a huge competitive edge on the market.

In many respects, Ajax is much like Apple. Ajax and Apple products exhibit flawless harmony between hardware and software custom built to be a perfect fit for each other. Both are also known for the attractive, carefully planned design of the entire product portfolio and software interface. At the same time (and unlike Apple today), Ajax assembles its devices at its own facilities, without delegating this process to subcontractors overseas. The company's owners believe that this is essential for keeping the bar high, especially when it comes to product reliability, which is crucially important in the security domain. When lives are hanging in the balance, any performance slip-ups simply cannot be tolerated.

So, what's in store for Ajax in the future? Oleksandr says the company doesn't plan to step off the gas pedal. "I see Ajax Systems as a multinational company that would be comparable to Apple in the area of security and smart homes. It would supply brilliant products to protect people and make their lives better by all possible means."

Why "Ajax"? Oleksandr Konotopskyi says it's part practicality, part art. "I'm a big fan of Greek mythology, so I remembered a hero with a resounding name — Ajax. I thought it was interesting that his name started with the first letter, A, and finished with one of the last ones, X. Besides, this name plays into our hands, too. We're always at the top of any alphabetical lists."

"OS Malevich was built on the idea of simplicity. We set a condition — new functionalities should not complicate the system or slow down the development process. So as not to stray off course, we gave a special code name to this project: Malevich, in honor of Kazimir Malevich, a remarkable Kyiv-born artist whose **Black Square** *exemplifies a brilliant idea based on infinite simplicity.*

"There was an idea of using Linux, but we turned it down right away because it was not protected enough from potential hacks. On top of that, it was highly recommended to use a real-time operating system in the development process, and Linux was not like that (real-time systems are used in devices and infrastructure facilities where a response to some event should be triggered within a strict time frame; that's why real-time OS are used in elevators, car brakes, and ballistic missiles).

"To develop OS Malevich, we had to change everything: the architecture, approach to programming, code style standards, workflow management, and programming environment. But as a result, we got a feature-packed OS that supports advanced cloud communication protocols over several channels and manages a network of hundreds of radio devices. It can simultaneously send alarm messages over IP channels, dial telephones, and send SMS. Attack-proof, it supports automation devices and has all the capabilities necessary for a professional security system."

When the full-scale invasion of Ukraine began, Ajax Systems opened a factory in a safer region of Ukraine. The company created new job positions there and relocated many employees. Ajax's regular shipments to partners were resumed in three weeks and the whole production cycle in five weeks. All this against the backdrop logistics issues and bombings.

Despite all the challenges of 2022, Ajax Systems managed to increase the company's revenue by 35% and their capacity by more than twice compared to the previous year.

↗
**Jeweller wireless devices
by Ajax Systems**

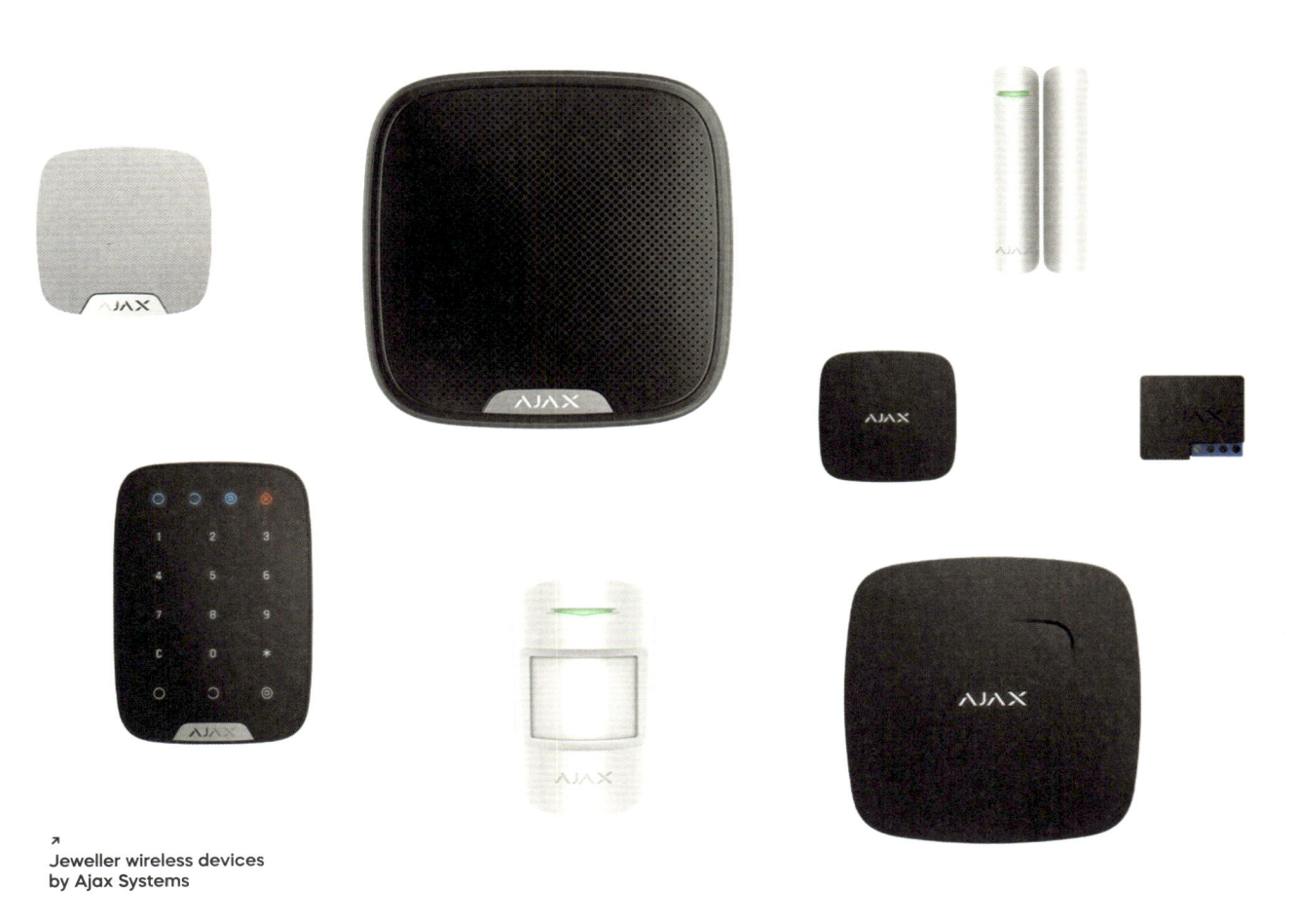

The company opened its first international factory in Turkey, continued to expand its product line, and enter new markets. Today, it unites over 2,700 employees and remains the industry's top player, even in a challenging, war environment.

> *"Work hard and always believe in yourself. Sounds simple, but actually not that many people in the world can do this. And remember — David always beats Goliath. So, be smart and fast. Don't be afraid of 'the big boys.' As they say, 'Amateurs built Noah's Ark, professionals built the Titanic.'"*
>
> *— Oleksandr Konotopskyi*

Reface: One Selfie Is All You Need

← Reface has seven founders and was launched in 2018 in Kyiv. Reface's mission is to empower everyone to create amazing content with AI.

Reface is the Ukrainian company that has been putting generative AI tools in the hands of ordinary mobile users since 2018 — and it all started with a selfie. Beginning with face-swap technology and growing to include other AI video editing tools, Reface's family of apps elevates the content game with generative AI-powered products straight from a pocket device.

Reface products empower people to push the limits of content creation via artificial intelligence. Before projects like ChatGPT or DALL-E were more than a rumor, Reface had already topped App Store rankings on two continents with its first AI-powered mobile application. What was behind this remarkable rise, and why has the app remained wildly popular?

Reface app simplified and democratized face-swap technology, allowing users to create fun videos with a single selfie.

Lessons in machine learning and the viral effect

Reface's story starts in 2011, when the company's future founders — Roman Mogylnyi, Oles Petriv, and Yaroslav Boiko — met as university students. They shared a passion for emerging technologies and began experimenting with machine learning (ML) long before it was a household term. Because ML allows computers to automatically learn from past data without explicit programming, the trio started training with text semantics (after all, this is how language acquisition naturally happens in humans). Before long, they had progressed from language to visuals and set up their own ML studio in Kyiv. Soon, they were partnering with Hollywood post-production studios.

Over the next few years, they ideated, practiced, and grew. At this stage they grew their expertise in computer vision, natural language processing, and 2D to 3D conversion. A big break happened in 2018 when a film studio client approached the team with an unusual request: they wanted to replace the faces of some of the actors on-screen. At first, the team agreed to take it on as an experiment. Modifying facial features was arduous work, but one day, a breakthrough came.

"This process could be sped up so it took just a few seconds instead of several weeks," remembers co-founder Oles Petriv. *"That's when our face-swap journey started."*

↗
Several dozen world celebrities used Reface app as a creative and entertaining way to communicate with their audience

It was a time dedicated to research and product development. During 2018 and 2019, the company expanded from three to seven co-founders — Ivan Altsybieiev, Dima Shvets, Den Dmytrenko, and Kyle Sygyda joined Roman, Oles, and Yaroslav — leading a full-fledged team.

202

Soon, the Reflect app was born. This app gave users a superpower: with one photo and a few swipes, anyone could transform themselves into the celebrity of their choice, selected from the app's curated photo library.

The company's debut app got a boost from celebrities themselves, who started sharing hilarious face-swapped selfies on social media. In 2019, Elon Musk blew up Twitter with a photo of himself digitally morphed with actor Dwayne Johnson. For Reface, it was a validation that the app was on the right track.

With just a bit of help, the app was soon promoting itself. As more users shared face swaps via Reflect, the trend spread like wildfire. It became clear to the owners that their product was something people definitely wanted.

Reface's next step was bringing face-swap technology to video. Doublicat, later renamed Reface, was rolled out in January 2020. In just one month, it was downloaded 100,000 times. By August, Reface was at the top of the US App Store. Before the year was out, the company had attracted investments totaling $5.5 million and Google had named it one of the best apps of the year. Within fourteen months of rebranding, the Reface app had reached 100 million downloads. (It took TikTok thirty-one months to do the same.) With backing from angel investor Andreessen Horowitz (a16z), Reface earned a new distinction: a Ukrainian startup able to compete with the best of the Silicon Valley.

From a one-app startup to a multi-product company

In the midst of an AI-revolution, Reface remains faithful to its original mission — to empower content creation. Anton Volovyk, co-CEO at Reface, calls this an ambitious goal in the rapidly shifting content creation market.

"The number of creators, tools, formats, and content distribution platforms is growing daily," he says. *"But we still believe that our calling is to democratize AI technologies — so ordinary smartphone owners can use them to meet their needs, whether it's making a funny birthday video or creating a unique avatar, and still have time for the bigger stuff in life."*

← One of the latest releases, the Restyle app gives any content dozens of different visual styles in one click.

Restyle app is one of the newest Reface products to do this. Essentially a magic wand for content creation, Restyle is the world's first video-to-video AI generator built for a mobile phone. With Restyle, a smartphone user anywhere in the world can reproduce the visual language of professional production studios on their own. The app allows users to transform photos and live-action videos into dozens of striking animation and illustration styles like anime, cyberpunk aesthetic, and even Impressionism.

Restyle has already breached the Top-10 in US and Latin America App Stores. Viral TikTok trends like "See yourself as an anime character" and "You as a doll" have brought the app's influence to tens of thousands of users worldwide.

Reface beyond the App Store

The company boasts several impressive collaborations with popular brands and A-list celebrities, among them international pop sensation Miley Cyrus.

"It was our first partnership program with an artist," Ivan Altsybieiev recalls of the 2020 partnership. *"Miley Cyrus's agents messaged us directly on Instagram. Imagine our social manager's face!" The result went viral.*

Millions of her fans got the chance to star in Miley's "Midnight Sky" music video with the help of Reface. Anyone could swap their own face in for Miley's face. (Miley herself played along by replacing her face with her godmother Dolly Parton's face.) The collaboration project won Reface recognition by *Billboard* and the company's name exploded.

Other high-profile partnerships include with Amazon Prime Video on the 2020 promo campaign for *Borat Subsequent Moviefilm*. The project gave anyone the chance to become Borat before the movie hit the screens. More than 750,000 fans shared their Borat "refaces" in less than five days.

But it was Kyiv-based fashion brand KSENIASCHNAIDER that introduced wider business use cases for Reface's technology. The brand, known for its signature recycled denim and collaborative drop with Adidas, faced closed runways for spring 2021 due to COVID lockdown restrictions. They came to Reface for help.

Reface teamed up with KSENIASCHNAIDER to produce a digital runway show and debut the Spring/Summer 2021 collection. From their homes, with large gatherings prohibited in most of Europe, viewers watched more than twenty models walk for KSENIASCHNAIDER. In reality, only three models took part in the fashion show — the "live" models they saw were actually digitally generated by Reface.

In January 2023, Reface's latest collaboration with BMW and creative agency Elastique put 15,000 visitors to the BMW Welt facility at the center of their own interactive spaceship. To create a memorable and futuristic experience for the launch of BMW's new vehicle model, Reface developed a high-resolution face-swap technology that allowed visitors to project themselves behind the steering wheel onto a 42-square meter 3D LED screen. The technology processes five video pieces simultaneously, replacing the default face with the user's face in a photorealistic look.

←
Reface & Elastique for BMW Mega Me campaign. Instead of watching a boring and impersonal ad, visitors could see themselves behind the steering wheel.

↑
Reface team is constantly experimenting with ML technologies and has a vision to establish an outstanding generative AI company from Ukraine.

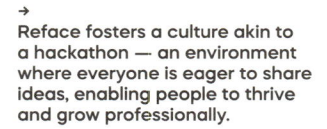

→
Reface fosters a culture akin to a hackathon — an environment where everyone is eager to share ideas, enabling people to thrive and grow professionally.

What lies ahead?

Are Reface apps a fad, another ephemeral mobile download that's here today and forgotten tomorrow? The answer to that question (it's "no") lies in the technology stack.

Behind each of Reface's apps, a team of engineers leverages advanced technologies to create surprising and entertaining new GenAI tools. Generative AI has been an incredible trend since the second half of 2022. BCG forecasts that by 2025, the industry will reach $60 billion and make up 30% of the entire AI market.

"This part of AI is designed to increase people's creativity; we only need to develop tools to unlock its potential," says co-CEO Anton Volovyk. *"Reface is in a solid position to become a global leader in building generative AI products."*

So, what are Reface's developers planning to do in the future? How are they going to use their technology? Most startups like Reface — hyped-up, fast-growing — either sell to larger companies or fizzle out as the viral effect fades away. Reface's founders avoided this fate by communicating to investors a clear vision of what comes next.

"Our vision is to establish an outstanding generative AI company from Ukraine," says Ivan, co-CEO at Reface. *"Our goal is to create many great products with generative AI. We strive to be a true GenAI player. Our focus lies in building user-friendly, scalable, and efficient products accessible to everyone. We want to be one of the best in this area and lead with our strengths in making AI technology easy to use."*

Oles Petriv insists that GenAI content creation tools are the beginning of the third major stage in the history of content development, after the invention of the printing press and then of computer games. "In our opinion, the next stage would be to combine the pattern of interactivity with content that's already available. That's what we call content recontextualization." The Reface team is working hard to have their technology enable this revolution in content. The future, as they see it, is one where anyone can be a creator.

"Ultimately, our goal is to build a large, successful company from Ukraine that makes a positive impact on people's lives," says Anton. *"This is the vision that drives us forward, and we are determined to make it a reality."*

As of 2023, the Reface company operates a portfolio of four GenAI products, including its original hit, the Reface face-swap app.

Jooble: Matching Talent and Companies Since 2006

← Iryna Paliienko, CEO of Jooble Direct to Employer; Jooble's founders Roman Prokofiev and Yevhen Sobakaryov; and Dmytro Gryn, CEO of Jooble Job Aggregator

When Jooble's founders Roman Prokofiev and Yevhen Sobakaryov launched their product in 2006, hardly anyone knew what a jobs board was. Instead, job openings were mainly shared via word of mouth or physical advertisements placed in high-traffic locations. But people did know Google. That's why when Roman and Yevhen took their product to the market, they chose a description anyone could understand, even if the product itself was unfamiliar: "The Google for job hunting."

Today, Jooble scours over 140,000 company websites, social media pages, and competitors to post comprehensive job listings for millions of job hunters in 67 countries. Jooble's success in making the job search fast and convenient on both ends of the hiring process is what keeps the site a favorite of job seekers and hiring managers worldwide.

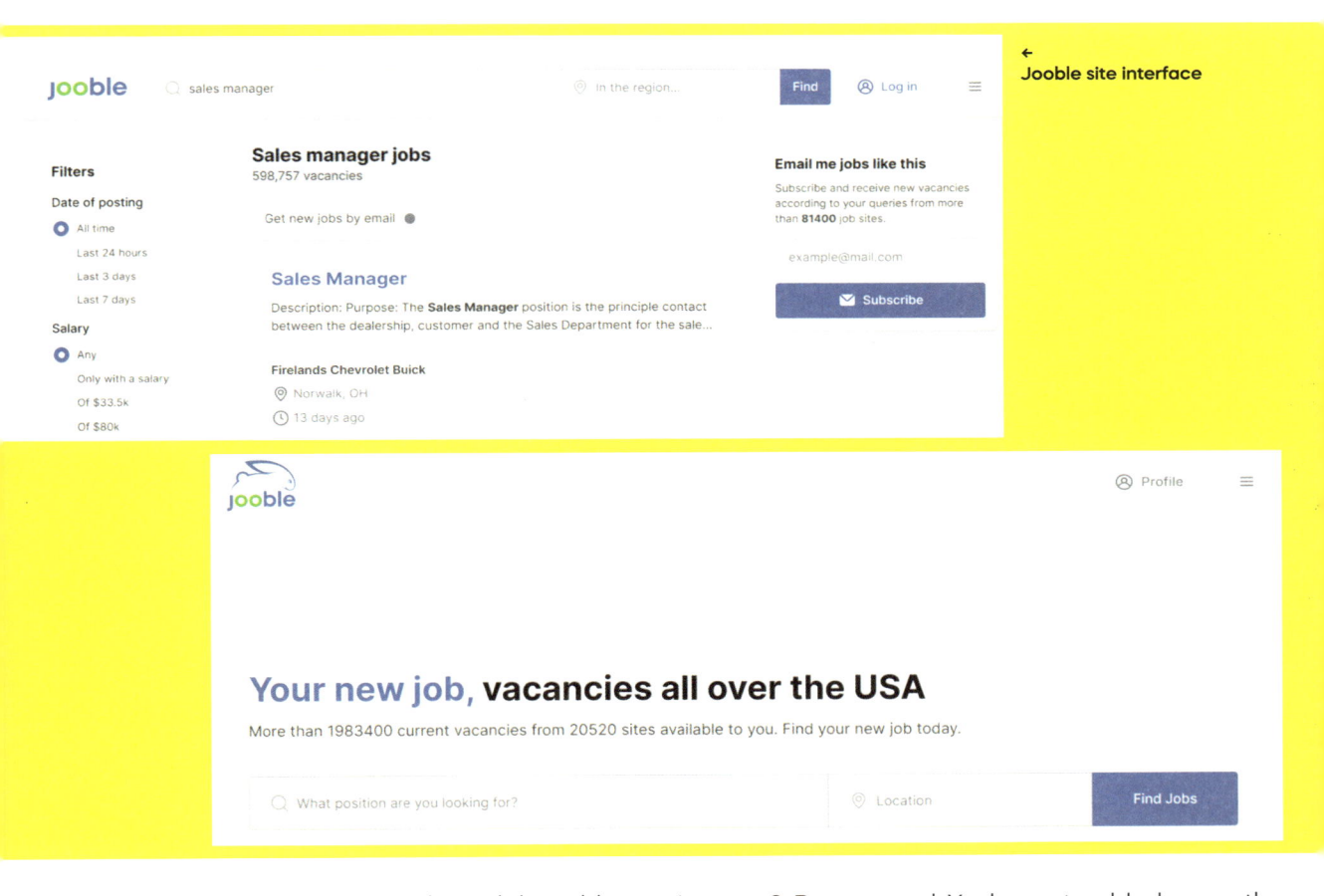

← Jooble site interface

So, how did Jooble get its start? Roman and Yevhen, stumbled upon the idea one afternoon over lunch at a cafe in downtown Kyiv. Friends since high school, they were both bright university students at Igor Sikorsky Kyiv Polytechnic Institute and talented programmers. And they were both bored by their studies.

The two friends had already been working for some time; Yevhen was selling CRM systems for Microsoft and Roman owned a successful pharmaceutical software startup, TeamSoft (today, 60% of medical representatives in Ukraine have used the platform). But itching for a new project, they tossed around

Mobile version
of Jooble's site

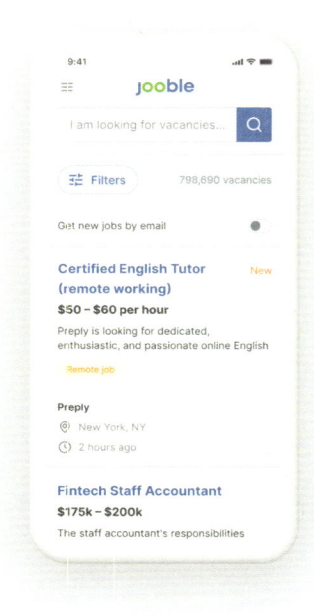

ideas between each other for a content aggregation platform that would challenge their coding skills. First, they thought of a news aggregation site but, inspired by the difficulties Roman was having in hiring talent for his company, ultimately decided on jobs listings as the best idea. They discussed the idea of the project right there and then, and six months later, Jooble was ready for its first visitors.

Young and enthusiastic, they got down to developing their product without conducting any market research. They didn't even check if there were any international competitors, as their solution was initially targeted at the local, Ukrainian market.

In retrospect, Roman says, "It was a serious mistake. You should always study someone else's experience and analyze what they did right and what they did wrong. On the other hand, though, you shouldn't dig into someone else's experience too deeply because you might give up on your own idea altogether."

Roman and Yevhen decided to attract their first job seekers through the most traditional type of advertising — ads pasted on the walls of university buildings. But luck was on their side, and they started attracting users even before the first ads went up on the wall. When Roman came to pick up the first batch of printed ads, a print shop employee told him that his daughter had found a job through Jooble. "I vividly remember that day. 'Wow!' I thought. 'No ads, nothing, and there's feedback already.'"

The printed ads did their job well, so Roman and Yevhen decided to spend $5,000 to place similar ads in the metro. Traffic on their website jumped from an average of 300 visitors a day to 3,000. Over the next couple of years, though, it just wouldn't get any larger. The project did pay off the investment, but the founders went through tough times. They were wondering if it made any sense to continue their business. By that time, Roman had invested $50,000 and wasn't eager to see it spent for nothing.

Luckily, the project started to make money, thanks largely to the pair's patience and determination to see the project through. But as soon as Jooble got firmly established, it had to withstand another test that brought about a pivotal decision. In 2008, the financial crisis broke out, profits dropped, and the company's prospects on the Ukrainian market became uncertain. "The market dried up in some two months, and our model collapsed," Roman said, remembering those days.

"We realized that there were two options — either we could close it all down and fire everyone, keeping just one person to support servers remotely, or we could try to enter other markets."

They went for the second option. International expansion started with Russian-language markets of Russia, Kazakhstan, and Belarus, as they were the most familiar to the company's founders. Jooble gained a foothold there quite fast. In 2009, after their initial international success, they decided to start serially expanding. Dmytro Gryn, CEO of Jooble Job Aggregator, says that expansion to international markets required both strategic and technical expertise.

"We broke down our product launch in new countries into a series of steps... We had to understand each country's geography and administrative structure — regions, towns, zip codes. This was the foundation of the job search process. We also had to learn to aggregate jobs and work with the local language. To do that, we used an in-house text processing library based on a range of stemming algorithms, primarily Porter stemmers." Stemmers are programs and algorithms that reduce words and descriptions to a common form. Put simply, they cut different word forms into a common root, and then search algorithms work with it.

The Jooble team also looked at the number of internet users, the internet penetration rate, and the GDP of the countries whose markets seemed large enough. Once a decision to expand was made, Jooble would be up and running in the chosen country in just six months.

"We had a schedule printed out on an Excel spreadsheet where it was marked that every two weeks our product was supposed to launch in two new countries," Dmytro remembers.

→
Photo of Jooble office in Kyiv, Ukraine

Since development costs were low, the biggest expenses were salaries, plus $200 to localize the interface for each new country. Jooble made the largest profit from the dynamic Western European and North American markets, where employment opportunities are constantly growing.

Roman recalls how the team set itself a firm deadline to launch Jooble in English-speaking countries by the end of 2009, "no matter what." When New Year's Eve rolled around, Roman said, "No one was leaving the office. Everyone was waiting to see the English-language version rollout. Once it was done, we all drank champagne and only then hurried home."

By 2010, it became evident that Jooble had made it as an international project. It had growth prospects, and the founders knew well how that growth could be sustained. TeamSoft, Roman's old company, didn't have any capacity for scaling up, so Roman decided to sell it — even though it was profitable — and focus on Jooble instead. Today, his old company is still up and running under a new name, Proxima Research.

When asked to give budding IT entrepreneurs some tips, Roman, just like many of his experienced colleagues, advises thinking globally.

"The Ukrainian market is too small to build a $100 million-dollar company. There's just a handful of domains — apart from e-commerce and fintech — where you could develop a business of this scale. Products that became world-famous, like Skype, were targeted at the international market right from the start, since their local distribution markets were too small.

No matter where the startuppers come from — Ukraine, the United States, or China — I always recommend them to have a global perspective and think of how they're going to scale up their product. I mean not just geographical presence. It's best to 'bite' as much as possible from the get-go."

How did Jooble manage to rise so high? One of the key success factors of any search engine are high-quality, relevant search results. To generate highly relevant results, Jooble uses machine learning algorithms.

Dmytro says that gathering data to train the ML algorithms has been a priority, "Since day one... the more data you have, the more conclusions you can make. If we see that one man working as, let's say, a brick mason, is interested in carpenter jobs, we can't really make some ultimate conclusion." But if the same correlation is repeated on a scale of 50,000, Jooble's algorithms can draw conclusions and offer job seekers more options. Sometimes, "options they didn't even think about."

Another success factor is speed. Jooble saves applicants and employers time — just take a look at its sleek interface.

"We designed a digital recruiter that recommends applicants to employers if they match the job requirements and have their profiles properly filled out. Another goal of ours is to speed up the process of posting jobs and filling out profiles. Job matching should be as fast as possible," Iryna Paliienko, CEO of Jooble Direct to Employer, says.

According to Jooble's founders, "the transition from a money-driven to a mission-driven company" in 2014 was another milestone in the company's history. Yevhen says that, in hindsight, Jooble's initial model "was of little value to job seekers." This despite running a successful business in 50 countries. "Our aggregator was functioning in the mode of a transitory business model. We drove traffic from Google and, using SEO, sold it to our clients. It was arbitration... Our goal was to earn as much money as possible."

Like many startups, it was only after the initial bills were paid that the company could start setting mission-based goals. "Two thousand and fourteen was like a test for our company because we realized that we were not that much interested in business as a money-maker only. We started to search for meaning in what we'd been doing, and it occurred to us that we'd like to not just make money but also to develop a product that helps applicants."

The company's goal now is to change the way applicants look for work and how employers find talent. Roman says that they are seeing many applicants for whom looking for work is "painful, stressful, and just terrible. And I see this not only in Ukraine but everywhere. Our dream is to change it, to create an ideal world where every person across their lifetime would be able to find work matching their aspirations and giving them a feeling of self-worth. This is our global plan. Not the end goal, but a vision of where we're going."

After the Russian invasion of Ukraine in 2022, the site ceased operating in Russia and Belarus. In the same year, Jooble and European companies launched the Give a Job for UA project, helping Ukrainian refugees with employment. Moreover, in 2022 Jooble created Jooble Venture Lab to grow an ecosystem of products that help people find work and fulfil themselves professionally. Also, Jooble launched the Jooble Job Search mobile app on iOS and Android.

Anyone visiting Jooble's website cannot help but notice its signature blue-and-white rabbit. "Why a rabbit?" many visitors might be wondering. Hardly anyone associates the job search with rabbits. And they're right. The story of the rabbit goes back to the founder's student days.

One early morning, Yevhen returned to his dorm room holding a white plush bunny. Roman was surprised. "Sobakaryov was a serious guy. To see him with a bunny rabbit stuffed animal was like seeing me in a tutu. 'I was at a birthday party,' he said. 'And I really liked this bunny! Let's make a logo with it!'"

The next day, Yevhen continued to push for the bunny logo. He argued that many successful IT companies and projects had animals as a mascot. Linux had a penguin; Mozilla had a fox, and Apple had Clarus the dogcow. Eventually, Roman gave in. "We thought that an image like that would be easier to remember than just a name. Associative memory is more effective. So, it was decided that the bunny would be there."

They commissioned a designer to draw a rabbit, but they felt that early sketches were too cartoonish. They adjusted the terms of reference, and the designer sent a new logo right away. "It was spot-on," says Roman. "It's this bunny that you can still see on our logo today. We didn't even experiment with the color. The designer drew a blue bunny, and we left it like that. We haven't changed it even once."

Appendix

Bibliography

«Головна справа життя академіка Глушкова» [Academic Glushkov's "Main Business of Life"]. UA Computing. https://uacomputing.com/stories/ogas/.

Bregis, E., Kuzima, A., & Revich, Y. (Eds.). (2016). *История информационных технологий в СССР. Знаменитые проекты: компьютеры, связь, микроэлектроника.* [History of Information Technologies in the USSR. Famous Projects: Computers, Communications, Microelectronics]. Knim.

Derkach, Vitaliy. *Академик В.М.Глушков – пионер кибернетики [Academician Glushkov, a Pioneer of Cybernetics].* Kyiv: Юниор, 2003.

"Development of Computer Science and Technology in Ukraine. Brief History." European Virtual Computer Museum. http://www.icfcst.kiev.ua/MUSEUM/museum.html.

Dolgov, Vyacheslav. *Анатолий Иванович Китов — пионер кибернетики, информатики и автоматизированных систем управления [Anatoly Ivanovich Kitov: A Pioneer of Cybernetics, Computer Science and Automated Control Systems].* Moscow: КОС.ИНФ, 2010.

Ehrenburg, I. (1990). *Люди, годы, жизнь [People, Years, Life].* Soviet Writer.

Энциклопедический справочник «Киев» [Encyclopedia "Kiev"]. Kyiv: Главная Редакция Украинской Советской Энциклопедии, 1986.

«Эврика» (сборники-ежегодники) [Eureka (Collections-Yearbooks)]. Moscow: Молодая гвардия, 1965–1980.

Genis, Alexander, & Weil, Peter. *60-е. Мир советского человека [The 60s: The World of the Soviet Man].* Moscow: АСТ, 1998.

Glushkov, Victor. *Введение в АСУ [Introduction to Automated Management Systems].* Kyiv: Техника, 1974.

Glushkov, Victor. *Введение в кибернетику [Introduction to Cybernetics].* Kyiv: Academy of Sciences of the Ukrainian SSR, 1964.

Glushkov, Victor. (Ed.). *Энциклопедия кибернетики [Encyclopedia of Cybernetics].* Kyiv: Главная Редакция Украинской Советской Энциклопедии, 1974.

Glushkov, Victor, Amosov, Nikolay. & Et. Al. *Энциклопедия кибернетики [Encyclopedia of Cybernetics]*. Kyiv: Редакция Украинской Советской Энциклопедии, 1975.

Glushkov, Victor, & Valakh, V. Y. *Что такое ОГАС? [What is OGAS?]*. Moscow: наука, 1981.

Glushkova, T. *После победы [After the Victory]*. Moscow: Ладога-100, 2002.

Большая советская энциклопедия (БСЭ) в 30 томах [Great Soviet Encyclopedia in 30 Volumes] (3rd Edition). Moscow: Советская энциклопедия, 1970–1978.

Ihor Sikorsky Kyiv Polytechnic Institute. (n.d.). "Вісник НТУУ 'КПІ.' Інформатика, управління та обчислювальна техніка" ["Bulletin of NTUU 'KPI.' Informatics, Management and Computer Engineering"]. Kyiv. https://kpi.ua/web_it-visnyk.

"История Создания Компьютеров 'Микро-80', 'Радио-86РК' и 'Микроша.'" ["The Story of the Invention of the 'Mikro-80' and 'Mikrosha' Computers."] История создания компьютеров "микро-80", "радио-86рк" и "микроша," 2011. https://zxbyte.ru/history.htm.

"Капітонова Юлія Володимирівна" ["Kapitonova Yuliya Volodymyrivna"]. Wikipedia. https://uk.wikipedia.org/wiki/Капітонова_Юлія_Володимирівна.

Keep, John. "Recent Writing on Stalin's Gulag: An Overview." *Crime, History, and Society* 1, no 2. Varia. 1997: p. 91-112. https://doi.org/10.4000/chs.1006.

Kolesnikov, A. *Попытка словаря. Семидесятые и ранее [Attempting a Dictionary: The Seventies and Before]*. Moscow: Рипол-классик, 2010.

Lelchuk, V. *Индустриализация СССР: история, опыт, проблемы [Industrialization of the USSR: History, Experiences, Problems]*. Moscow: Издательство политической литературы ЦК КПСС (Политиздат), 1984.

Макроэкономические модели и принципы построения ОГАС [Macroeconomic Models and Principles of OGAS Construction]. Moscow: Статистика, 1975.

Maksimovich, G. V. *Беседы с академиком В. Глушковым [Conversations with the Academic V. Glushkov]*. Moscow: Молодая гвардия, 1978.

Malynovskyi, Borys. *Академик В. Глушков: Страницы жизни и творчества [Academician V. Glushkov: Pages of Life and Creativity]*. Kyiv: Наукова думка, 1993.

Malynovskyi, Borys. *Документальная трилогия [Documentary Trilogy]*. Kyiv: Горобець, 2011.

Malynovskyi, Borys. *Очерки по истории компьютерной науки и техники в Украине [Essays on the History of Computer Science and Technology in Ukraine]*. Kyiv: Феникс, 1998.

Malynovskyi, Borys. *Первая отечественная ЭВМ и её создатели (к 40-летию ввода МЭСМ в регулярную эксплуатацию) [The First Domestic Computer and Its Creators (on the occasion of the 40th anniversary of the introduction of the MESM into regular operation)]*. Moscow: Наука, 1992.

Malynovskyi, Borys & Glushkov, Victor. *Золотые вехи истории компьютерной науки и техники в Украине [Golden Milestones in the History of Computer Science and Technology in Ukraine]*. Kyiv: Відкритий міжнародний університет розвитку людини "Україна". 2003.

Malynovskyi, Borys. *История вычислительной техники в лицах [The History of Computer Technology in People]*. Moscow: КИТ, 1995.

Malynovskyi, Borys. *Хранить вечно [Keep Forever]*. Kyiv: Горобець, 2007.

Malynovskyi, Borys. *Маленькие рассказы о больших учёных [Little Stories About Great Scientists]*. Kyiv: Горобець, 2013.

Malynovskyi, Borys. *История вычислительной техники в лицах [Pioneers of Soviet Computing]*. Kyiv: фирма "КИТ," 1995.

"NedoPC.Org." nedoPC.org. "Трезвый критический взгляд на 86РК." ["A sober critical look at 86RK."] Accessed 2023. http://www.nedopc.org/forum/viewtopic.php?f=93&t=17237.

Osipchuk, Ihor. «Академик Глушков от души смеялся, посмотрев пародии о своем институте кибернетики» ["Academician Glushkov Laughed Heartily After Watching Parodies About His Institute of Cybernetics"]. OGAS. January 9, 2014. http://ogas.kiev.ua/ua/perspective/akademyk-glushkov-ot-dushy-smeyalsya-posmotrev-parodyy-o-svoem-ynstytute-kybernetyky-766.

Parfyonov, L. M. *Наша эра [Our Era]*. Moscow: Колибри, 1961–1970.

Peters, Benjamin. *How Not to Network a Nation: The Uneasy History of the Soviet Internet*. Cambridge: The MIT Press, 2017.

Ponomarenko, L., & Riznyk, O. *Київ. Короткий топонімічний довідник [Kyiv: A Short Toponymic Reference Book]*. Kyiv: Павлім, 2003.

Rosenthal, M., & Yudin, P. (Eds.). *Краткий философский словарь под редакцией [A Short Philosophical Dictionary]*. Moscow: Издательство политической литературы ЦК КПСС (Политиздат), 1954.

Sedov, E. A. *Занимательно об электронике [Interesting Things About Electronics]*. Moscow: Молодая гвардия, 1967.

Serbyn, Roman and Krawchenko, Bohdan. *Famine in Ukraine*. Edmonton: Canadian Institute of Ukrainian Studies, University of Alberta, 1986.

«Шкабара Катерина Олексіївна». ["Shkabara Kateryna Oleksiivna"]. Wikipedia. https://uk.wikipedia.org/wiki/Шкабара_Катерина_Олексіївна.

Shvets, Marta. "Kateryna Yushchenko: The Programmer Who Changed the World". Stem Is Fem. https://stemisfem.org/en/ushchenko.

Sobolev, S. L., Kitov, A. I., & Lyapunov, A. A. *Основные черты кибернетики [The Main Features of Cybernetics]*, Вопросы философии, no. 4 (1955).

«Ющенко Катерина Михайлівна». ["Kateryna Mykhaylivna Yushchenko"]. Wikipedia. https://uk.wikipedia.org/wiki/Ющенко_Катерина_Михайлівна.

«Вера Глушкова». ["Vera Glushkova"]. DataArt IT Museum. https://museum.dataart.com/ru/narratives/vera-glushkova-ch-2.

Verstyuk, V., Dziuba, O., & Reprintsev, V. *Україна від найдавніших часів до сьогодення. Хронологічний довідник [Ukraine from Ancient Times to the Present — A Chronological Guide]*. Kyiv: Наукова думка, 1995.

"Віктор Глушков. Життя та творчість. Ілюстрації. Література." ["Victor Glushkov: Life and Creativity. Illustrations. Literature."]. International Charity Foundation for History and Development of Computer Science and Technique. http://www.icfcst.kiev.ua/MUSEUM/GL_HALL2/photos_u.html.

Weiner, N. *Кибернетика, или управление и связь в животном и машине [Cybernetics, Or Control and Communication in Animal and Machine]*. Moscow: Советское радио, 1958.

Williams, N., Waller, P., & Rowell, J. *Chronology of the 20th Century*. Abingdon: Helicon, 1996.

Acknowledgements

The authors would like to express their deep gratitude to the following people for their invaluable research contributions:

Borys Malynovskyi
Pavlo Shevelo
Vira Glushkova
Georgy Gimel'farb
Viktor Perchuk
Volodymyr Petrukhin
Vira Bigdan
Vladyslav Husev
Volodymyr Puida
Maryna Tarasova
Oleksandr Kovalenko
Dmytro Cherepanov
Mykola Shcherbyna
Oles Maygutyak
Oleksandr Rybak
Stefan Mashkevych
Volodymyr Grynchuk
Mykola Shvets
Ben Peters
Lev Zaslavsky
Volodymyr Gurovych
Tamara Malashok
Ihor Karev
Sergii Prokofiev
Dmytro Yashanov
Lyubov Borovkova
Vasyl Pihorovych
Kateryna Horiaieva
Yuliia Shakhnovska
Maryna Rohova
Sergii Legusha
Sergii Frolov
Dmytro Sergeev
Oleksii Pedosenko
Yaroslav Azhnyuk
Max Lytvyn
Alex Shevchenko
Dmytro Lider
Denys Zhadanov
Oleksandr Kosovan
Oleksandr Konotopskyi
Roman Mogylnyi
Oles Petriv
Yaroslav Boiko
Den Dmytrenko
Kyle Sygyda
Ivan Altsybieiev
Dima Shvets
Roman Prokofiev
Yevhen Sobakarev
Dmytro Gryn
Iryna Paliienko

The MacPaw team extends additional thanks to:

Natalia Pysarevska, Director of State Polytechnic Museum at Igor Sikorsky Kyiv Polytechnic Institute, and Kostyantyn Antonenko, Staff Scientist, for providing access to archival and historical materials.

Mykola Antonyuk, Maryna Boichenko, Anna Kalabska, and Kateryna Zlenko for legal advice.

Amina Yepisheva, Anastasiia Fed-Titova, Mariia Sibirtseva, Anastasiia Bilynska, Andriy Klen, Brad Wells, Kateryna Chystiakova, Tetiana Repetska, Maria Henyk, and Yelyzaveta Grenda for assistance in preparing company chapters.

Anthony Bartaway for assistance with editing.

Alona Korolova, Ivan Taranenko, Yuriy Fedorenko, Mykhailo Alvares, Oleksandr Koval, Pavlo Romaniuk, Iryna Mazurak, and Yaroslav Valentyi for creating the website and promotional materials

Yevheniia Palash, Tetiana Bronytska, Daria Perekosova, Mykhailo Berveno, Anna Manukhina, Nadiia Kolesnykova, and Yurii Husynskyi of MacPaw team for assistance with the project.

Volodymyr Nevzorov
Victoria Ugryumova

Innovation in Isolation.
The Story of Ukrainian IT
from the 1940s to the
Present.

Paper: Imitlin Tela Neve 125 g/m², cardboard 2,5 mm, Munken Pure 130 g/m²
Size: 84×90/16
Typeface: Mazzard H, Space Mono
Edition of 2000 copies
Offset printing
Printing: "Publish Pro" Ltd.
publishpro.com.ua

Copyright for the book:
MacPaw, Kyiv, 2024

Copyright for the book materials:
This book may not be used in whole or
in part in any form without the written
consent of MacPaw.

Publisher: ist publishing
istpublishing.org

Certificate of inclusion of the subject
of publishing in the State Register of
Publishers, Manufacturers and Distributors
of Publishing Products DK № 5289 dated
18.02.2017

MacPaw